D1295736

Introduction to Radio Propagation for Fixed and Mobile Communications

The Artech House Mobile Communications Series

John Walker, Series Editor

Advanced Technology for Road Transport: IVHS and ATT, Ian Catling, editor

An Introduction to GSM, Siegmund M. Redl, Matthias K. Weber, Malcolm W. Oliphant

CDMA for Wireless Personal Communications, Ramjee Prasad

Cellular Digital Packet Data, Muthuthamby Sreetharan and Rajiv Kumar

Cellular Mobile Systems Engineering, Saleh Faruque

Cellular Radio: Analog and Digital Systems, Asha Mehrotra

Cellular Radio Systems, D. M. Balston, R. C. V. Macario, editors

Cellular Radio: Performance Engineering, Asha Mehrotra

Digital Beamforming in Wireless Communications, John Litva, Titus Kwok-Yeung Lo

GSM System Engineering, Asha Mehrotra

Introduction to Radio Propagation for Fixed and Mobile Communications, John Doble

Land-Mobile Radio System Engineering, Garry C. Hess

Mobile Communications in the U.S. and Europe: Regulation, Technology, and Markets, Michael Paetsch

Mobile Antenna Systems Handbook, K. Fujimoto, J. R. James

Mobile Data Communications Systems, Peter Wong, David Britland

Mobile Information Systems, John Walker, editor

Personal Communications Networks, Alan David Hadden

RF and Microwave Circuit Design for Wireless Communications, Lawrence E. Larson, editor

Smart Highways, Smart Cars, Richard Whelan

Spread Spectrum CDMA Systems for Wireless Communications, Savo G. Glisic and Branka Vucetic

Transport in Europe, Christian Gerondeau

Understanding GPS: Principles and Applications, Elliott D. Kaplan, editor

Vehicle Location and Navigation Systems, Yilin Zhao

Wireless Communications in Developing Countries: Cellular and Satellite Systems, Rachael E. Schwartz

Wireless Communications for Intelligent Transportation Systems, Scott D. Elliott, Daniel J. Dailey

Wireless Data Networking, Nathan J. Muller

Wireless: The Revolution in Personal Telecommunications, Ira Brodsky

For a complete listing of *The Artech House Telecommunications Library*, turn to the back of this book.

Introduction to Radio Propagation for Fixed and Mobile Communications

John Doble

Artech House
Boston • London

Library of Congress Cataloging-in-Publication Data
Doble, John.
Introduction to radio propagation for fixed and mobile communications / John Doble.
p. cm.
Includes bibliographical references and index.
ISBN 0-89006-529-2 (alk. paper)
1. Mobile communication systems. 2. Microwave communication systems. 3. Radio wave propagation. I. Title.
TK6570.M6D63 1996
621.384'11—dc20 96-26605
CIP

British Library Cataloguing in Publication Data
Doble, John
Introduction to radio propagation for fixed and mobile communications
1. Radio wave propagation
I. Title
621.3'84'11

ISBN 0-89006-529-2

Cover design by Kara Munroe-Brown

685 Canton Street
Norwood, MA 02062

International Standard Book Number: 0-89006-529-2
Library of Congress Catalog Card Number: 96-26605

10 9 8 7 6 5 4 3 2

To Jill—
Without your help and constant encouragement,
this book would not have seen the light of day.

Contents

ACKNOWLEDGMENTS

I wish to thank the following for so readily giving their permission to use the material indicated below:

- BT Labs for the multipath fading pattern of Figure 2.1 and the comparison between diversity and nondiversity fading patterns of Figure 3.4, originating from studies carried out on the Swaffham-Mendlesham experimental link. Also for the set of system signatures, Figures 6.5 to 6.7, the ray diagrams in Figures 2.5 and 2.6, and the geographic factors in Figure 6.3 for use with the U.K. multipath prediction model.
- The Radiocommunications Agency for the results of mobile studies funded by them and carried out at Liverpool and Bradford universities.
- The management committee of COST 231 for permission to use their developments of path loss-prediction models for use in the 1.8-GHz band.
- Dr. John Norbury from the Rutherford Appleton Laboratory, for the curves relating line-to-point rain rates in Figure 5.2.
- The International Telecommunications Union (ITU) for numerous diagrams and extracts from reports and recommendations, as specifically indicated in the text.

Please note the following about the material from ITU texts:

1. These texts are reproduced with prior authorization with ITU as copyright holder.
2. The sole responsibility for selecting extracts for reproduction lies with the beneficiary of this authorization alone and can in no way be attributed to the ITU.

3. The complete volumes of the ITU extracts can be obtained from:

International Telecommunications Union
General Secretariat, Sales and Marketing Service
Place des Nations
CH-1211 Geneva 20 (Switzerland)

Telephone: +41 22 730 51 11 Telex: 421 000 uit ch
Telegram: ITU GENEVE Fax: +41 22 730 51 94
X.400: S=Sales; P=itu; A=Arcom; C=ch
Internet: Sales@itu.ch

INTRODUCTION

▾▾▾

This book is specifically directed toward giving the reader a basic insight into the propagation effects that he or she will encounter when working in the areas of fixed-link (microwave line-of-sight) and mobile radio systems. We start by defining what is meant by the term *propagation*.

Put into the simplest terms, propagation studies encompass everything that happens to a radio signal from the time it leaves the transmitter output port until it arrives at the receiver input port, including any effects experienced at the interface between the antennas used and the transmission medium. Even this is not the whole story, as you must also consider the various equipment used, ranging from narrow-band analog to wide-band digital, and how this equipment behaves under stress.

In fixed links, it is the structure variability of the atmosphere and the consequential effects on the transmission of the microwave signal, with the attenuation from heavy rainfall, that is our chief interest. Nevertheless, you must keep in mind that a knowledge of what is happening within the transmission medium tells you very little unless it relates to the "total system."

In mobile systems, it is the particular way in which we use the transmission medium, coupled with the antenna characteristics, that introduces performance limits, and we will find that this leads to considerable differences between fixed-link and mobile systems in certain areas of transmission.

The book is divided into two sections. Part I, comprising Chapters 1 to 7, concerns fixed links. Part II, Chapters 8 to 12, deals with mobile systems and explains how the major differences between fixed links and mobile arise. We briefly describe the subject matter in each section below.

PART I FIXED-LINK SYSTEMS

This section starts with a discussion of the radio refractive index of Earth's atmosphere. It addresses why it varies with height (a property that leads to the transmission of microwave signals beyond the geometric horizon) and explains how to perform refractivity measurements. It then shows how the change in refractivity with height can deviate from normal and explains the causes of these deviations and some of their effects on the transmission of microwave signals.

This leads to a deeper investigation (Ch. 2) into the subject of the fading of signals in microwave systems, in which two statistically independent components of fading are present. One of the components is *frequency independent* and the other, by reason of the mechanism by which it is generated, is *frequency selective.* The effects of these fading components on analog and digital systems, together with antenna performance, are then examined, concluding that analog systems are reasonably insensitive to the type of fading present; whereas digital systems are much more sensitive to the frequency-selective component. One important property of microwave antennas is degradation in the presence of frequency-selective fading, which leads to the development of high-performance antennas to reduce its severity.

Diversity-reception methods have been developed to reduce the severity of this fading, and Chapter 3 investigates the pros and cons of the various approaches used. We emphasize that although the main aim of diversity is to reduce the severity of fading, it also introduces a reduction in various forms of distortion and interference.

Deep fading can also be a factor in the introduction of several forms of interference, and Chapter 4 emphasizes the precautions you must take in the design stage of the equipment development, as well as the planning of the route itself, to minimize problems of this type.

Radio-refractive index perturbations are not the only cause of fading. Chapter 5 discusses the topic of signal attenuation caused by rainfall along the signal path, details two approaches to prediction of fade depth, and lists the available sources of rain-rate data. An example comparing the results from the two prediction techniques is included. Chapter 5 also considers the signal attenuation brought about by molecular absorption by water vapor and oxygen.

Chapter 6 covers the important topic of system performance targets and outage modeling. The outage prediction in analog systems requires only a simple technique, but digital systems demand more complex approaches, each requiring access to databases of system performance related to fade depth. Thus, although the principles behind the alternative approaches are common knowledge, you cannot use these methods unless the required databases are available.

The final chapter in Part I deals with various aspects of system planning techniques for ensuring that the system meets internationally agreed performance objectives. It also presents link budget; two approaches to the fading prediction; and a worked planning exercise for an analog link, which uses much of what you learned in this section and includes the determination of antenna heights at each end of a link. The need to take account of water anywhere along the path and methods of reducing its impact are also investigated.

PART II MOBILE SYSTEMS

The first chapter in this second section, Chapter 8, discusses characteristics of various mobile services and the frequency bands used. It then deals with basic propagation topics at VHF, UHF, and the lower microwave frequencies; the mobile environment; and the role of reflection and diffraction paths in the propagation of mobile signals.

Chapter 9 covers the important topic of path loss predictions at VHF, UHF, and microwave frequencies in rural, urban, and suburban areas. It explains why the path loss experienced is greater than that encountered in fixed-link systems, and discusses practical measurements of path loss and delay, and details how results from the latter should best be presented.

We discuss cellular systems in Chapter 10, including the principal characteristics of analog and digital equipment and the definitions of cell types from propagation rather than from traffic aspects. The benefits of cell splitting and sectorization for meeting the increased demands on the system as you move from rural to urban situations are detailed with comments on using diversity and equalization.

Chapter 11 considers the topic of communication into and within buildings; defines penetration loss and height gain; and discusses measurement techniques. We examine prediction models and their accuracy and include some examples of anomalous measurements results.

The final chapter discusses the various propagation aspects of fixed-link and mobile that might, on casual inspection, be expected to have a degree of commonality. They turn out to be very different, however, and Chapter 12 examines the underlying factors that cause these differences and their impact.

It is the aim of this book to enable the reader to gain a good oversight of a variety of propagation topics, to obtain an understanding of the problems associated with developing propagation models, and to give a firm basis on which to build a thorough understanding of fixed-link and mobile transmission. The information within this book is based either on first-hand practical experience or on knowledge gained while working in a number of international organizations.

NOTES

For our purposes, we consider microwave frequencies to cover a 900-MHz to 60-GHz range. The lower limit is usually 1 GHz, but with cellular radio systems straddling this frequency, it is best to use 900 MHz as a starting point. Similarly, the boundary between microwaves and millimetrics is normally set around 30 GHz; however, the nature of transmission around 60 GHz makes it of great interest to the mobile community, so we have also extended the upper limit.

The International Radio Consultative Committee (CCIR) is now known as the Radiocommunication Sector of the ITU. We refer to it by its former title in this book since it is reports issued under this name that contain much important information.

PART I

Fixed-Link Systems

CHAPTER 1

RADIO REFRACTIVE INDEX

1.1 INTRODUCTION

In this chapter, we start with a section on the background of microwave radio links that includes an examination of the nature of the radio refractive index (RRI) of Earth's atmosphere. We will discover that it has a smooth variation with height above Earth's surface under normal conditions and understand why this is so important to the fundamental viability of microwave radio systems.

We also discuss the various ways to perform refractivity measurements and learn why refractivity can sometimes change greatly over a small height interval. The particular significance of these events on the transmission path, leading to undesirable effects on the microwave signal, forms the subject matter of the first half of the final section of the chapter, which concludes with an explanation of the various meteorological mechanisms that lead to their formation.

1.2 BACKGROUND TO MICROWAVE RADIO LINKS

The individual links in microwave radio networks are often referred to as being line-of-sight (LOS) links, so let us start by posing two questions:

1. What does line-of-sight mean?

2. What practical and regulatory factors limit the path length of a microwave link?

In answering the first question we must understand that the term *line-of-sight* does not mean straight-line transmission. The structure of the Earth's atmosphere is such that its refractive index varies with height, and this property causes significant curvature of the transmission path. Under median conditions, the refractive index of the atmosphere decreases with height in an exponential manner, but, over the height range of interest to system planners, we can consider the characteristic to be linear.

A very basic representation of the transmission path is shown in Figure 1.1, in which the atmosphere is represented by a concentric shell structure in which each shell has a refractive index that is uniform over its height range but decreases with the distance of the shell from Earth's surface. Hence, the transmission of a radio signal through such a structure is in the form of a number of linear sections with a downward angular change occurring at each interface between layers of different refractive indices.

As the thickness of each layer is reduced, we eventually arrive at the true situation in which the change of refractive index is a continuous process, and the transmission path follows a smooth curve with a radius of approximately four times that of the Earth. Thus, the microwave signal is transmitted beyond the geometrical horizon, Figure 1.2, and permits the transmission of signals over reasonably long distances without excessively high antenna-support structures.

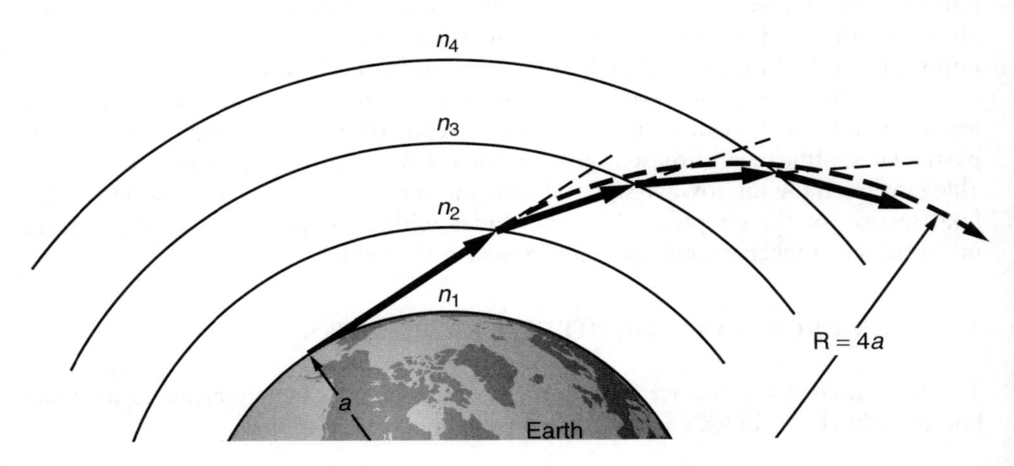

FIGURE 1.1 Concentric shell model of Earth's atmosphere.

FIGURE 1.2 Geometric and microwave radio horizons.

It would help if, before answering the second question, we have a short introduction to microwave antennas to define some of the terms you will meet in this and following chapters.

The *gain* of an antenna is defined as the ratio, usually expressed in decibels, of the power required at the input to a reference antenna to the power supplied to the input of the given antenna to produce, in a given direction, the same field strength. The reference antenna usually adopted at microwave frequencies is an *isotropic radiator* (one that radiates equally in all directions).

The most commonly used antenna in the microwave part of the radio spectrum is the *parabolic reflector antenna* in which the reflector is illuminated either by a feed unit mounted at the focus (a front-fed antenna), or by a feed unit mounted such that it radiates through an aperture at the center of the reflector and illuminates a subreflector mounted at the focus (a *cassegrain antenna*). The beamwidth of such antennas is comparatively small, and so the field strength at a given distance from the antenna is much greater than that produced by an isotropic source—hence, the high gain of such antennas. For a given frequency, the larger the diameter of the antenna, expressed in wavelengths, the smaller the beamwidth and the higher the gain.

If a parabolic antenna were perfect, then power from it would radiate in one direction only. This situation, however, cannot be achieved in practice. One imperfection is that for whichever type is constructed, there is need for a support structure to locate a component at the focus. This structure intercepts a fraction of the transmitted power and scatters it into many directions. Also, the illumination of the reflector is a compromise between making full use of the reflector's area while trying to avoid energy spillover at the edge. Thus, there is some energy radiated in directions other than that required. A plot of radiated power against angular offset from the desired direction of transmission reveals a complex pattern, the main peak in the direction of transmission being known as the *main lobe,* with a number of other peaks known as *sidelobes.* These sidelobes are several tens of decibels lower than the main lobe, and can nevertheless cause problems under certain conditions.

The final term that requires explanation at this point is *cross polarization.* The energy can be radiated from the antenna with the electric field either horizontal or vertical. Whichever orientation you choose, there will be components of the other in the received signal from either imperfections in the feeders and/or the antenna, or from particular conditions within the transmission medium. The ratio between the two received components, expressed in decibels, is the *cross-polar discrimination* of the system. This value is around 45 dB in an ideal situation but degrades during adverse transmission conditions.

The existence of cross-polar discrimination has encouraged the practice of transmitting two carriers at the same frequency but on different polarizations to use the spectrum more efficiently. Considerable difficulties arise in this area, however, as we discuss later.

Turning to the second question, let us examine the variables that a microwave network planner might manipulate to determine the parameters of a link he or she wants to install. First, power sources are expensive, and the planner will thus go for the lowest powered source that enables the link to function effectively under the full range of expected transmission conditions. Second, the planner must bear in mind that the free-space-transmission loss equation contains terms of 20log d and 20log F, where d is hop length in kilometers and F is frequency in megahertz. (We will later find that further path losses from rain attenuation apply at the higher microwave frequencies.) The easy solution appears to opt for a relatively short hop length and/or as low a frequency as possible. The first of these, however, is economic suicide, since all the installation and provision costs relate to the terminal/repeater stations; whereas the intervening path, besides a fee from the licensing authority, does not cost in terms of hop length. (Practically, there is little possible manipulation of this variable because of the fixed locations between which the traffic originates.) The low operating frequency may well be a viable option, but it also has ramifications as we see below.

It appears on the face of things that since microwave reflector antennas are reasonably compact, the designer can overcome the basic problems by using high-gain antennas over as long a range as possible. Antenna gain, however, can be related to the reflector diameter expressed in wavelengths. Therefore, if a low frequency is selected, the antenna dimensions are larger, wind stresses on the structure are increased, and a more expensive support structure is required. Further, the longer the link's length, the higher the planner must mount the antenna to avoid diffraction losses over hilltops along the route. Note there is also a frequency-conscious term to determine the necessary clearance of the radio path over obstacles along the route, and this again rules against using lower frequency bands.

Within the United Kingdom networks, the vast majority of long-distance, high-capacity traffic is carried in the 4-, 6-, 8-, and 11-GHz bands, and the choice of frequencies is dictated by the network as a whole, with the need to avoid interference into other routes becoming the controlling factor. So the problem finally resolves itself by juggling the power source, feeder losses, and antenna gains to ensure that the receiver is driven at its optimal level under median operating conditions. Note that if the antenna gain demanded is too high, a very narrow beamwidth results, which in turn leads to signal losses under conditions of high winds or adverse propagation. A beamwidth of approximately 1 degree between the 3-dB points of the antenna is normal. For the regulatory aspect, the licensing requirements will almost certainly contain a reference to the maximum permitted effective indicated relative power (EIRP), so when juggling the above parameters, take care to avoid exceeding this figure.

The result of the restrictions on the various parameters is that hop lengths in the 40- to 50-km range is the norm within the United Kingdom for links operating in the 4- to 11-GHz bands. We will find later that above 11 GHz, the attenuation from precipitation becomes the predominant feature of planning, so that for links operating in the 19-GHz band, for instance, the upper limit of hop lengths is around 8 km.

The ability to transmit signals over significant distances makes inter-city communication by microwave-radio circuits economically viable. Because the costs are in the terminal/repeater equipment, the longer the radio sections used, the cheaper the overall systems. The structure of the atmosphere is variable, however, depending on the humidity, pressure, and temperature at any given point. It is this variability, with some associated undesirable effects on the microwave signal, that needs thorough understanding.

1.3 SEA LEVEL VALUE OF RRI AND ITS VARIATION WITH HEIGHT

The value of the atmospheric refractive index n at sea level varies with season and location. For the United Kingdom, the mean value is taken as 1.00034. It is very inconvenient to work with a variable that is so close to unity, and it is conventional to use N units, the relationship between the two being

$$N = (n - 1) \cdot 10^6 \tag{1.1}$$

This results in a mean sea level value of N in the United Kingdom of 340 units. The seasonal and location variation, on a global basis, amounts to ± a few tens of units [1]. N is strictly refractivity but is usually referred to as the RRI.

The proportions of the principal gases that constitute the atmosphere—nitrogen, oxygen, rare gases, and carbon dioxide—maintain a nearly constant interrelationship with increase in height; whereas the quantity of water vapor is extremely variable and becomes a very significant factor in the variability of RRI.

The main factors affecting RRI are pressure, temperature, and water vapor pressure, the relationship for frequencies in the radio portion of the spectrum being reasonably described by

$$N = (77.6 \times P/T) + (3.73 \times 10^5 \times e/T^2) \tag{1.2}$$

Where the pressure P is in millibars, the temperature T is in Kelvin, and the water vapor pressure e is in millibars [2]. The two terms involved in the relationship are often referred to as N_{dry} and N_{wet} respectively, and at any given temperature N_{dry} makes a relatively constant contribution of about 265 to N, and N_{wet} provides most of the variability.

1.4 THE MEASUREMENT OF RRI

To understand the variability of microwave transmission, it is important to be able to relate its behavior to the RRI profile within the bottom few hundred meters of the atmosphere and also to know how this profile varies with distance along the transmission path being investigated. There are various techniques for gathering the necessary data, and it is useful to understand the practical limitations of the various techniques.

1.4.1 Radiosondes

As part of the international weather data-collection program, most countries release radiosondes at a number of locations on a regular basis, normally midday and midnight. The information so obtained can indicate the behavior of the atmosphere for radio propagation studies. The sonde comprises instruments for measuring temperature, pressure, and humidity, together with the necessary electronics to radio the information back to the receiving station on the ground. (References for each of these variables are also contained within the package.) The arrangements for transmitting measured data plus reference signals back to the ground is such that transmissions of pressure data are made at intervals of 35m of ascent, that of humidity at every 25m, and temperature at every 10m. The time response of the sensors, coupled with the interval between data transmissions from the radiosonde, means that any fine detail of the refractivity gradients is lost, and the use of data collected by this means can only give a very crude feel for the refractivity profile.

1.4.2 Refractometers

Any serious investigation into the subject uses fast-response devices, either carried on an aircraft, carried on a balloon, or spaced apart up a high tower. Such instruments are based on the measurement of the change in resonant frequency in a cavity with partially open ends, due to a change in the refractive index of air being passed through it (usually by a fan). The results from such an instrument show that what is indicated as a fairly uniform reduction of RRI with height, as measured by a meteorological radiosonde, can in fact have a very detailed microstructure, some of the perturbations extending over as little as a meter interval of height. See Figure 1.3.

Clearly, any profile derived from balloon-carried or tower-mounted instruments can only provide a localized profile, resulting in sweeping assumptions if you assume the information is valid over a path of up to 50 km. The aircraft-mounted refractometer, however, can supply details of the RRI profile over a range of locations if the aircraft carries out a series of shallow climbs and dives as it progresses along the route of the radio path.

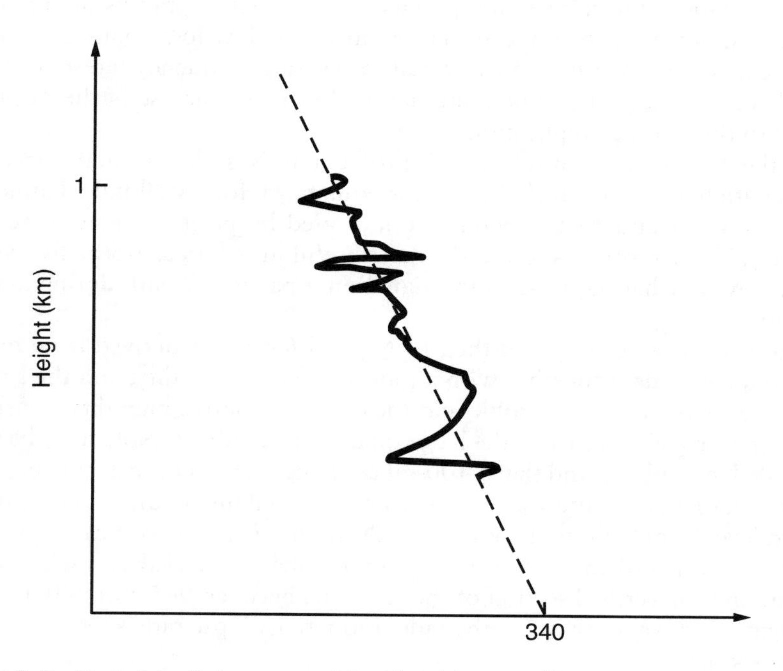

FIGURE 1.3 Typical detailed structure of the refractivity N-profile.

1.4.3 RRI Profiles

Although the RRI profile tends to an exponential form, assume that the lapse rate in a well-mixed atmosphere is linear within the first kilometer of height, with a slope of –40 N-units/km (known as the *clear-sky value*). Deviations from this norm, of course, are indicated by a change in slope. Within a complex situation, however, it is not easy to identify the degree and sense of any deviations from normal, so the N-profile is of little practical use.

There are two alternatives to overcome this problem. In the first, the modified refractivity or M-profile has a factor $h/r_0 \times 10^6$ (where h is the height above sea level and r_0 is Earth's radius) added to the N value at any given height. If one uses a value of 6,370 km for r_0, then at a height of h km:

$$M(h) = N(h) + 157 \times h \tag{1.3}$$

Therefore, $dM/dh = 0$ when $dN/dh = -157$ N-units/km and a plot of M against h is vertical for this critical value at which a phenomenon called *ducting* takes place. In a ducting situation, the refractivity gradient is such that the radius of curvature of the transmission path is the same as that of Earth, and low-loss, long-distance transmission occurs. Such a situation can result in severe cofrequency interference problems and is the subject of ongoing international studies. The use of the M-profile is restricted to this specific application.

In the second alternative, the K-profile removes the normal variation of N-values with height, so that dK/dh is approximately 0 for a well-mixed atmosphere and sub- or superrefractivity regions are indicated by positive or negative slopes, respectively. The K-profile is found the most useful in practice, especially when trying to determine what happens to the signal on a particular path during abnormal refractivity.

Figure 1.4 is an example of the M-, N-, and K-profiles derived from meteorological radiosonde data together with a plot of relative humidity, and the similarity between the shapes of the K-profile and the curve of relative humidity clearly demonstrates the strong influence of the latter on the RRI gradient. Note that the vertical axis is scaled in milibars, and that a 100-mbar change represents a 1-km height interval. This lack of an absolute scale arises from the variability of atmospheric pressure at ground level from flight to flight of the radiosonde. The near vertical section of the M-profile between 990 and 976 mbar represents a height interval over which ducting is present, and the vertical section of the K-profile between 965 and 940 mbar indicates a height interval over which the radio refractivity lapse rate is the median value of 40 units/km.

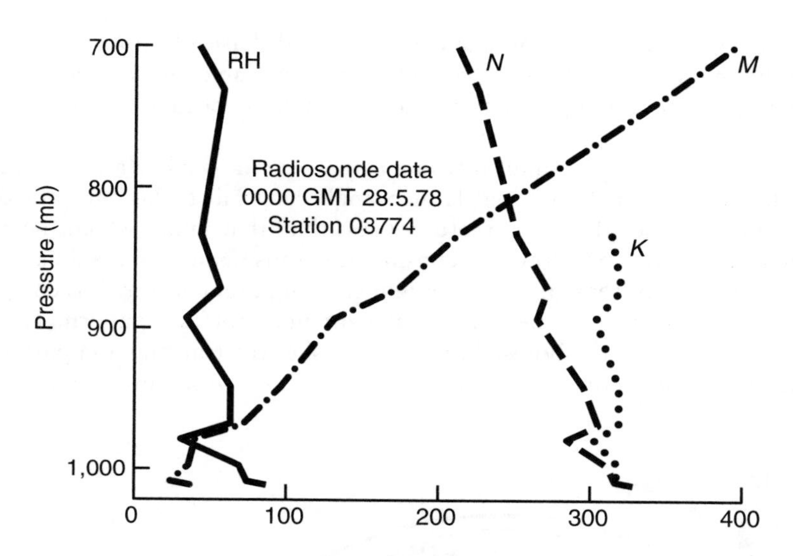

FIGURE 1.4 Profiles of *M*, *N*, *K*, and relative humidity.

1.5 DEVIATIONS FROM THE NORMAL LAPSE RATE

Although the normal lapse rate of RRI is around 40 *N*-units/km, there can be deviations from this situation. These deviations manifest themselves as height regio_ having lapse rates less than or greater than the clear-sky value from certain meteorological conditions. The height interval involved in such deviations can be a meter or less at one end of the scale, rising to a few hundred meters at the other extreme. However, because of the limited range of refractivity variation available (remember that there are approximately 340 units between the mean sea-level value and vacuo), the greater the height range encompassing the perturbation, the less the possible deviation from the median value of 40 *N*-units/km. Also note that a region of sub- or superrefraction cannot exist in isolation, there being a balanced system with the value of RRI returning to the normal value at some height outside the region affected by the particular meteorological condition responsible for the abnormal situation.

1.5.1 Subrefraction

The antennas on a microwave link are aligned during a period when the atmosphere is well mixed, so that any atmospheric layering is destroyed and a lapse rate of

40 *N*-units/km prevails. The antennas are adjusted, basically for maximum signal level, although you may have to consider other variables (discussed later). With this alignment, the antennas cater for the normally expected ray curvature from refraction.

If the radio path lies completely in a height region of subrefraction, then a single ray transmitted at the normal launch angle may fail to be intercepted by the receive antenna, since the ray curvature is less than that under which the antennas were aligned (Figure 1.5). However, consider the transmission as a fan-shaped group of rays, with those offset from the boresight carrying progressively less energy as the offset increases, because of the shape of the antenna radiation pattern. Thus, it is a ray transmitted below the boresight that establishes the transmission path resulting in an offset loss of energy. Note that since the path between antennas is reciprocal,

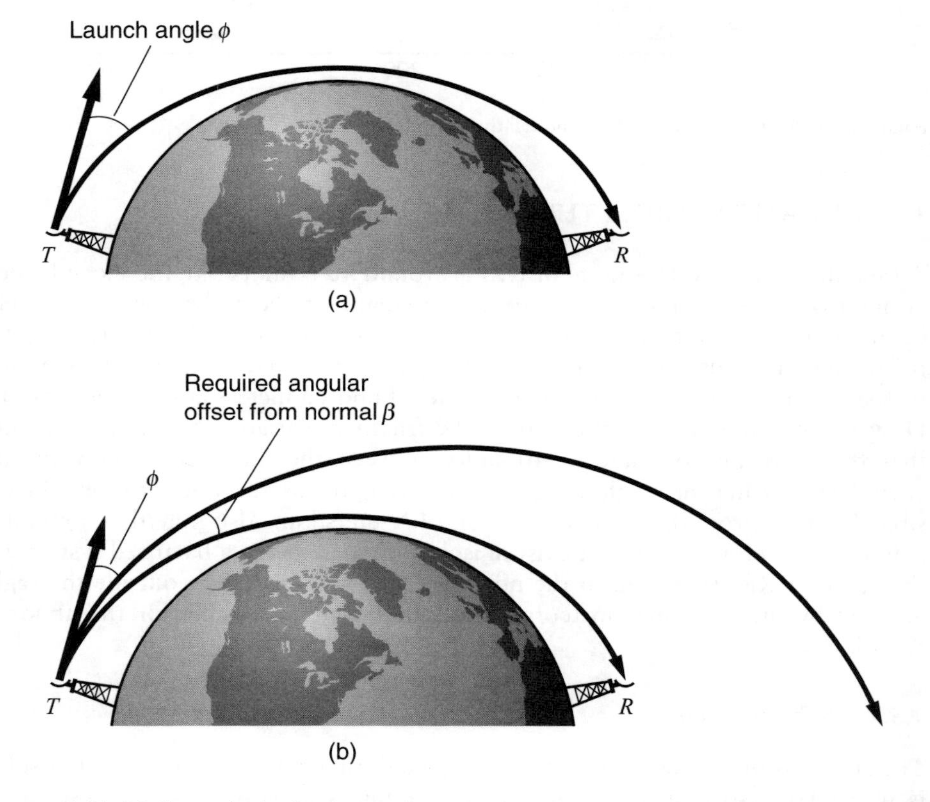

FIGURE 1.5 Median and subrefractive ray diagrams: (a) normal lapse rate, ray launched and received at angle ϕ, and (b) lapse rate less than 40 units/km, normal ray not intercepted by receive antenna). Communication is established via ray launched at angle β below the normal and hence suffers attenuation of both antennas due to main-lobe patterns.

offset loss is encountered at both antennas. (We will show later that this may not be the only type of signal degradation incurred under these conditions.)

1.5.2 Superrefraction

For this condition, an offset launch-receive angle again occurs, this time as compensation for the greater-than-normal ray curvature, again resulting in signal loss (Figure 1.6).

1.5.3 The Common Causes of the Lapse-Rate Deviations

There are four commonly recognized mechanisms that can modify the clear-sky RRI lapse rate as follows.

Radiation Nights

This is a very common situation in which a sunny day is followed by clear skies overnight. During the day, Earth absorbs heat from the sun, and the air temperature rises. After sunset, Earth radiates heat into space, and the surface temperature drops. Because Earth is a poor conductor of heat, this surface heat loss is not replaced from below, and consequently the air adjacent to Earth cools faster than the air higher up. We thus have a condition that is known as a *temperature inversion* (a linear decrease in air temperature of 1° K/100m of height being the usual situation), and the RRI profile no longer has a uniform lapse rate. Note that because of the better thermal conductivity of water, compared with that of Earth, this situation is not possible over the sea or other large expanses of water.

Advection Effects

These are common to many coastal regions in which a moderately dimensioned expanse of sea (50–250 km) separates land masses. As an example, consider the situation shown in Figure 1.7, in which the East Anglian region of the United Kingdom is separated from the European land mass by the North Sea, which is around 160 km wide at this point. If a region of high pressure is located over northern Germany, then the clockwise movement of air around the "high" causes warm, dry air over the land to be swept over the relatively cool, moist air above the sea, and once again we have a temperature inversion. This inversion is swept over East Anglia,

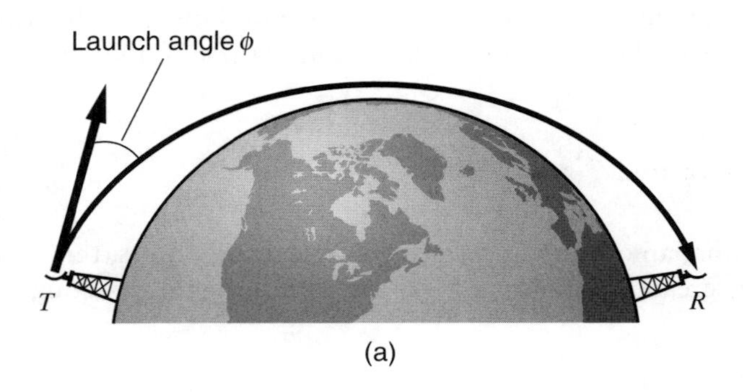

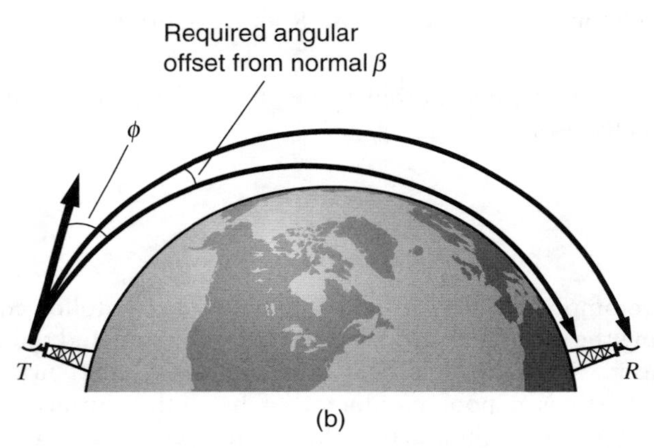

FIGURE 1.6 Median and superrefractive ray diagrams: (a) normal lapse rate, ray launched and received at angle ϕ, and (b) lapse rate greater than 40 units/km, normal ray not intercepted by receive antenna. Communication is established via ray launched at angle β above the normal and hence suffers attenuation at both antennas due to main-lobe patterns.

and will persist, as the system moves inland, until it encounters rising ground, causing vertical mixing of the atmosphere and the collapse of the inversion.

Subsidence

This effect is brought about by descending air in a high-pressure system becoming heated by compression. Having come from a high altitude, the air is relatively dry and can spread over the cooler, moist air below, once again forming a temperature inversion. Normally, the inversion established by this mechanism tends to stabilize at around 1 km above Earth's surface and does not affect transmission on microwave links. There are occasions, however, when, because of interaction between the subsidence event and an incoming frontal system, the layer descends to as low as 100m, giving rise to very severe effects indeed [3]. The full mechanism involved in events of this type is not yet fully understood. Studies in this area continue.

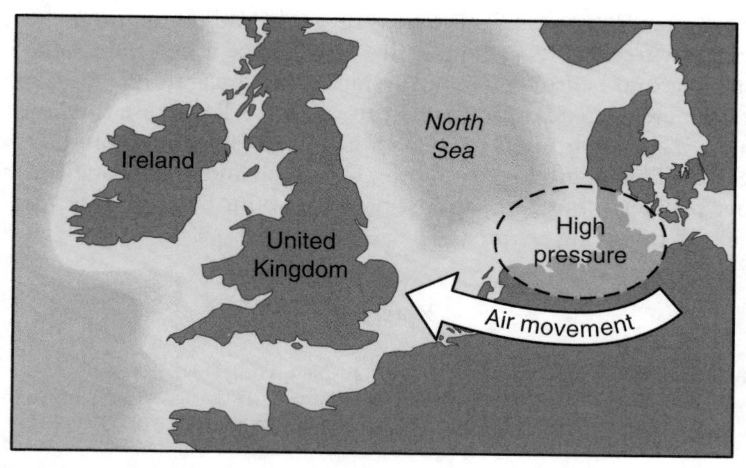

FIGURE 1.7 Meteorological basis of advection events.

Frontal Systems

A cold front approaching an area can cause the situation in which a wedge of cold air is driven in beneath the warmer air in front of it and causes a localized inversion. The speed at which these fronts move ensures that any propagation disturbances are short term.

These are the four main causes of abnormal lapse rates, but there can be other effects that may cause localized problems.

1.6 SUMMARY

The structure of Earth's atmosphere is such that its refractive index decreases with height above its surface. Although this decrease is an exponential function, for all practical purposes we can consider it linear over the height range of interest to microwave engineers. The atmospheric refractive index has a median sea-level value little removed from unity. It is more convenient to work with refractivity units that have a sea-level value of around 340 units and a lapse rate of 40 units/km under median conditions.

This refractivity variation with height results in the path of a microwave signal being bent down toward Earth's surface with a radius of approximately four times that of Earth, under median conditions, resulting in propagation of the signal beyond the geometric horizon. This extension of the radio horizon makes the installation of microwave circuits economically viable since it can reduce the number of repeater stations and hence the overall system cost significantly.

Certain meteorological conditions can cause the RRI lapse rate, over a limited height range, to change from its median value by a significant amount. One outcome is an increase in signal path loss from off-axis launch and arrival angles at the antennas. Further propagation degradations, brought about by lapse-rate variations, are discussed in the following chapter.

References

[1] Bean, B. R., and Dutton, E. J., *Radio Meteorology,* New York, Dover, 1966.

[2] CCIR, *Reports of the CCIR 1990, Annex to Volume 5,* ITU, Geneva, 1990, Recommendation 369–4, p. 41.

[3] Doble, J., "6 GHz Propagation Measurements Over a 51-km Path in the UK." *AGARD Conference on Terrestrial Propagation Characteristics in Modern Systems of Communications, Surveillance, Guidance, and Control,* Ottawa, October 1986, AGARD CP-407, pp. 20-1 to 20-8.

CHAPTER 2

FADING ON MICROWAVE SYSTEMS

2.1 INTRODUCTION

Fixed links, for a small proportion of time, suffer performance degradations caused by signal fading as a result of RRI anomalies. We discuss in this chapter the mechanisms by which this fading occurs, the identification of fading components, and the very different effects of fading on analog and digital system performance. Finally, we examine the effects of fading on certain important features of antenna performance and the influence of these effects on measures aimed at using the radio spectrum as efficiently as possible.

2.2 THE MECHANICS OF FADING

If, because of one of the meteorological conditions described in Chapter 1, atmospheric layering is present, then it is possible that energy from a transmitted signal may arrive at the receiver by more than one path. A direct signal path occurs as well as a delayed path brought about by, say, the abnormal refraction of an offset signal component within a superrefractive layer positioned above the direct signal path. When radio signals simultaneously follow two or more paths of differing physical length, between transmitter and receiver, these signals then suffer different transmission delays due to the finite propagation velocity involved. The vector combination of these signals at the receiver can range from destructive to additive, depending on the relative delays involved. This resultant variation of the observed signal level is known as *multipath fading*.

The term *multipath* is somewhat of a misnomer since, unlike HF circuits, we can identify the vast majority of the fading experienced on microwave links as being due to two-path situations [1]. Nevertheless, the term is generally used to describe fading from the simultaneous existence of more than one transmission path. However, the character of a typical fading event, (Figure 2.1) is revealed by monitoring the automatic gain control (AGC) circuit of the receiver (which controls the receiver gain during fading to maintain its output at the normal level). This shows that there is a slowly varying component—the *mean-depression* or *flat-fading* term, together with a much faster variation—the *frequency selective* term, which is superimposed upon it. It is only this second term which is described by the mechanism above, but the term *multipath* tends to be incorrectly attached to the overall fading pattern.

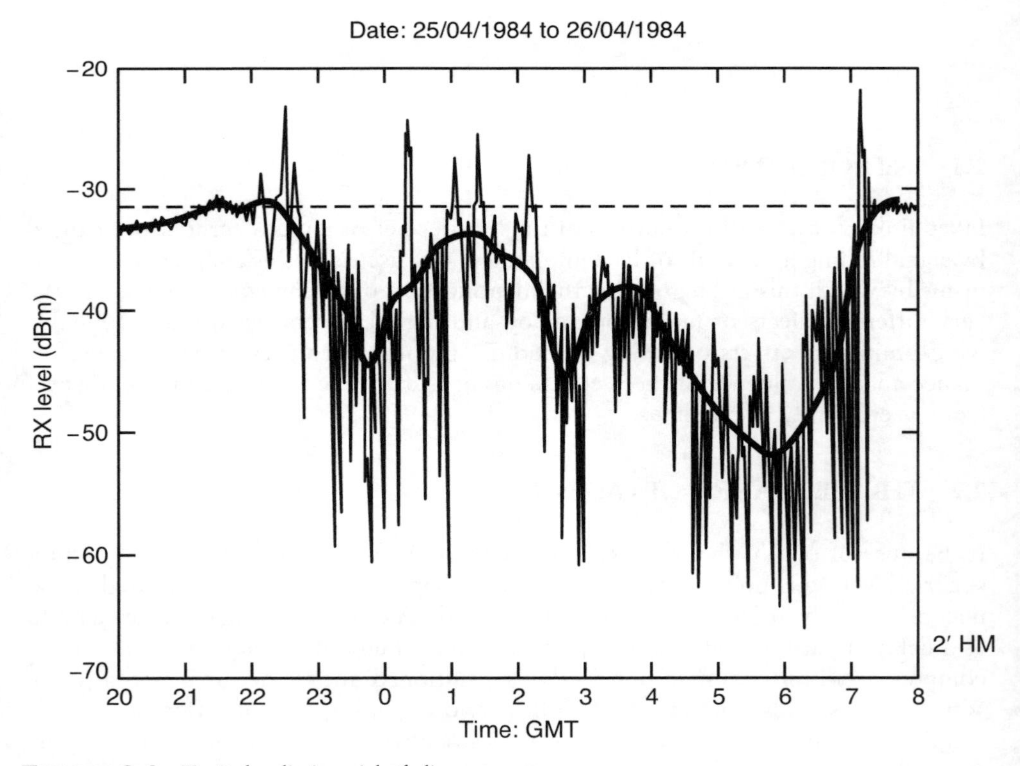

FIGURE 2.1 Typical radiation night fading event.

It is interesting to understand how the mislabeling of the combined fading came about. It arose from the fact that microwave links were, for many years, exclusively narrowband analog in nature. Such links were sensitive to signal-level loss, the presence of fading being revealed by increase of background noise as the receiver AGC system operated. Thus the operator was primarily interested in the percentage of time for which fading exceeded the threshold at which the link became unusable (system flat-fade margin), irrespective of the nature or cause of the fading. The more recent move toward broadband digital communication links, which are relatively insensitive to the mean-depression fading but do suffer errors from the distortion introduced into the system by the frequency-selective component, has aroused interest in determining the distributions of the two components of fading. Much work is currently being carried out in this area. One particular approach is to sweep a "time window" through narrowband fading records. Supposing fading records from an analog link were available with signal-level samples taken at one-second intervals. If the width of the time window was, say, one minute, then the mean-depression level during that minute would be the mean value of all the samples taken, and the level of frequency-selective fading at the midpoint of the window would be the difference between the level of the midpoint sample and the mean-depression level. Take care to establish a window width that produces meaningful results from the available data, however.

Let us now take a closer look at the two components of fading (which are statistically independent) and the effect they have on systems and antennas.

2.2.1 Multipath Fading

Since the true multipath fading is caused by the interaction of signals emanating from the same source, but suffers differing delays en route, it follows that the fading is frequency conscious since it is only at frequencies for which the differential delay between the two components is an odd number of half-wavelengths that we can observe the maximum fade depth. This leads to the alternative but more descriptive name for the fast-varying component of fading: *frequency-selective fading*. Let us take the basic case of a two-path situation in which there is no mean-depression fading present. Figure 2.2(a) represents a simplistic situation in which there is a direct path P_d and a path P_r involving lossless refraction from atmospheric layering. (For the sake of simplicity, normal ray curvature is ignored in this diagram.) The vector combination R of the two terms (Figure 2.2(b)) is defined by:

$$\begin{aligned} R^2 &= (P_d + P_r \cdot \text{Cos } \varnothing)^2 + (P_r \cdot \text{Sin } \varnothing)^2 \\ &= P_d^2 + P_r^2 + 2 \cdot P_d \cdot P_r \cdot \text{Cos } \varnothing \end{aligned} \tag{2.1}$$

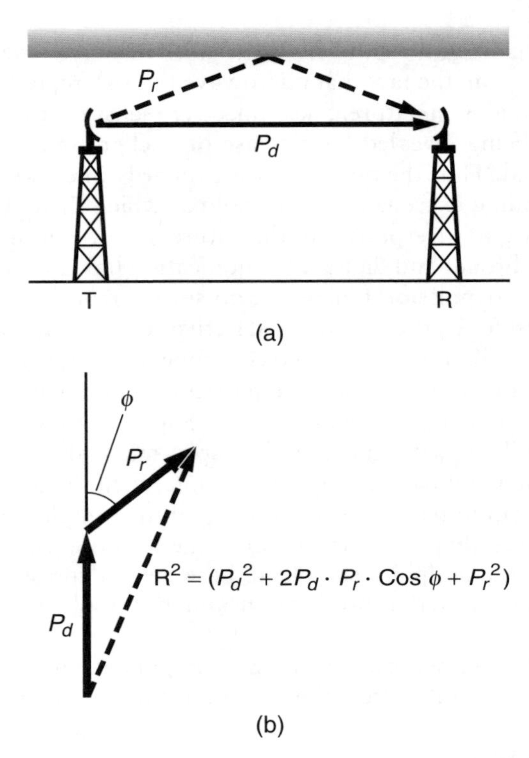

FIGURE 2.2 Multipath fading mechanism: (a) multipath geometry and (b) vector relationship of multipath components.

For $n = 1,2,3, \ldots$ then if $\varnothing = 2n\pi$ radians, there is a direct summation of the two signals; and if $\varnothing = (2n - 1)\pi$ radians, some cancellation occurs.

The characteristic frequency-selective form of two-path multipath is shown in Figure 2.3. The deepest part of the fades (usually referred to as *notches*) occurs at the frequencies (GHz) where τ, the relative delay between the two signal paths in *n*Sec, is a factor of $(2n - 1)/2$ of the periodic time of the signal.

In practice, the delayed path will usually be attenuated slightly, relative to the direct path, because of the antenna main-lobe discrimination at the offset launch/receive angle, and hence as the delay decreases (i.e., the offset angle is reduced), the discrimination is less and deeper notches occur. Understand as well that some situations can give rise to infinite notches, and the particular significance of such events is explained later when we discuss the effect of multipath fading on wideband systems.

The upper limit of multipath delays that can cause deep notches in a LOS system is usually considered to be around 6 ns [2]. This is because delays greater than

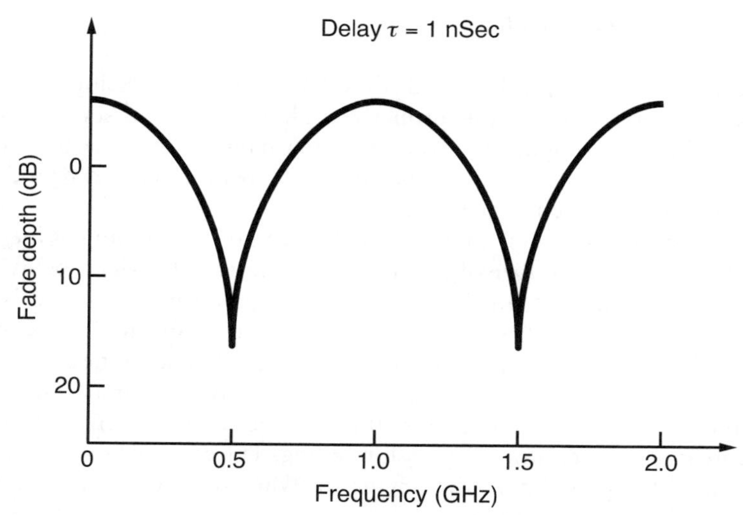

FIGURE 2.3 Repeat multipath fading pattern in the frequency domain

this only arise from atmospheric layering well removed from the direct signal path and therefore give rise to large offset launch and receive losses. Thus deep fading from near cancellation of the two signal components cannot occur under these conditions.

The time of day at which a multipath event occurs can give a strong clue as to the mechanism of the RRI event responsible. A radiation night event, for instance, requires some considerable time after sunset to allow Earth's surface to cool and to establish the temperature inversion. Once the atmospheric layering is present, it is very unlikely to break down before sunrise, when the heating of Earth's surface causes setup of convection currents and the atmosphere becomes well mixed. The fading pattern shown in Figure 2.1 is typical of a radiation night event. In contrast, an advection event, as seen in eastern England as a result of the mechanism indicated in Figure 1.7, occurs in the late afternoon, once the inversion forms and the layering drifts across the North Sea. In the worst scenario, an advection event can last through to the start of a radiation night, both being caused by the same high-pressure system, resulting in more than 12 hours of multipath activity.

There should be no particular reason why a subsidence event should occur at any particular time of the day. Nevertheless, examination of the occurrence of these indicates a tendency toward overnight situations. Why this should be is not clear, but it no doubt emphasizes our lack of detailed knowledge of the mechanisms involved.

2.2.2 Mean-Depression Fading

Unlike the multipath fading described above, the mean-depression component of fading has no frequency-conscious properties. Thus, if you had several antennas at the same height on a supporting structure, each having the same beamwidth but working in different bands, then a similar mean-depression pattern would be experienced by each of the systems involved.

In addition to loss of signal from offset launch and receive angles mentioned in Section 1.4, here are a number of other common causes described below with the aid of ray-tracing diagrams. In these diagrams, the transmitted signal is modeled as a group of rays angularly offset from each another by an 0.5-mrad interval (Figure 2.4). Each ray is attenuated relative to the center one by the shape of the antenna main-lobe pattern. In some cases, no rays are shown to arrive close to the receive antenna, but this does not mean that no power at all is received. It does, however, indicate quite severe mean-depression fading. The example ray-tracings should be compared with Figure 2.5(a), the tracing for when there is no layering present—known as the *clear-sky condition.*

(a) Figure 2.5(b) shows *defocusing* (also known as *beam-spreading*) of the antenna main-lobe pattern as a result of a complex RRI profile within the transmission path. In this diagram, the transmit antenna is within a region in which a superrefractive zone lies below one of subrefractivity. Rays launched close to but below the horizontal are bent downward and arrive

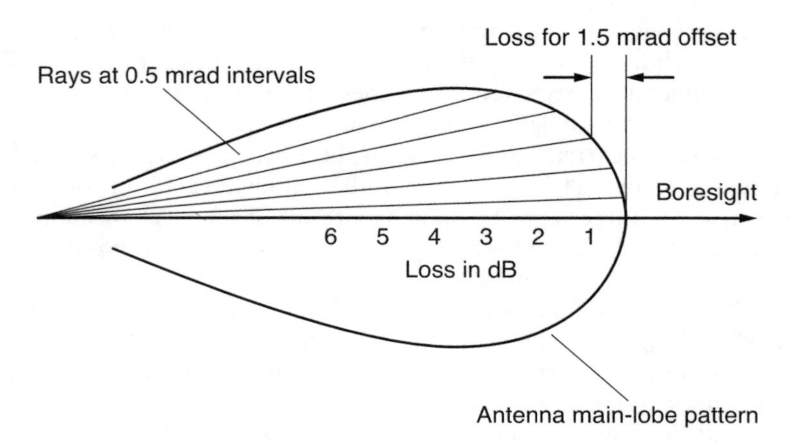

FIGURE 2.4 Modeling of radiated signal by rays offset at 0.5 mrad intervals.

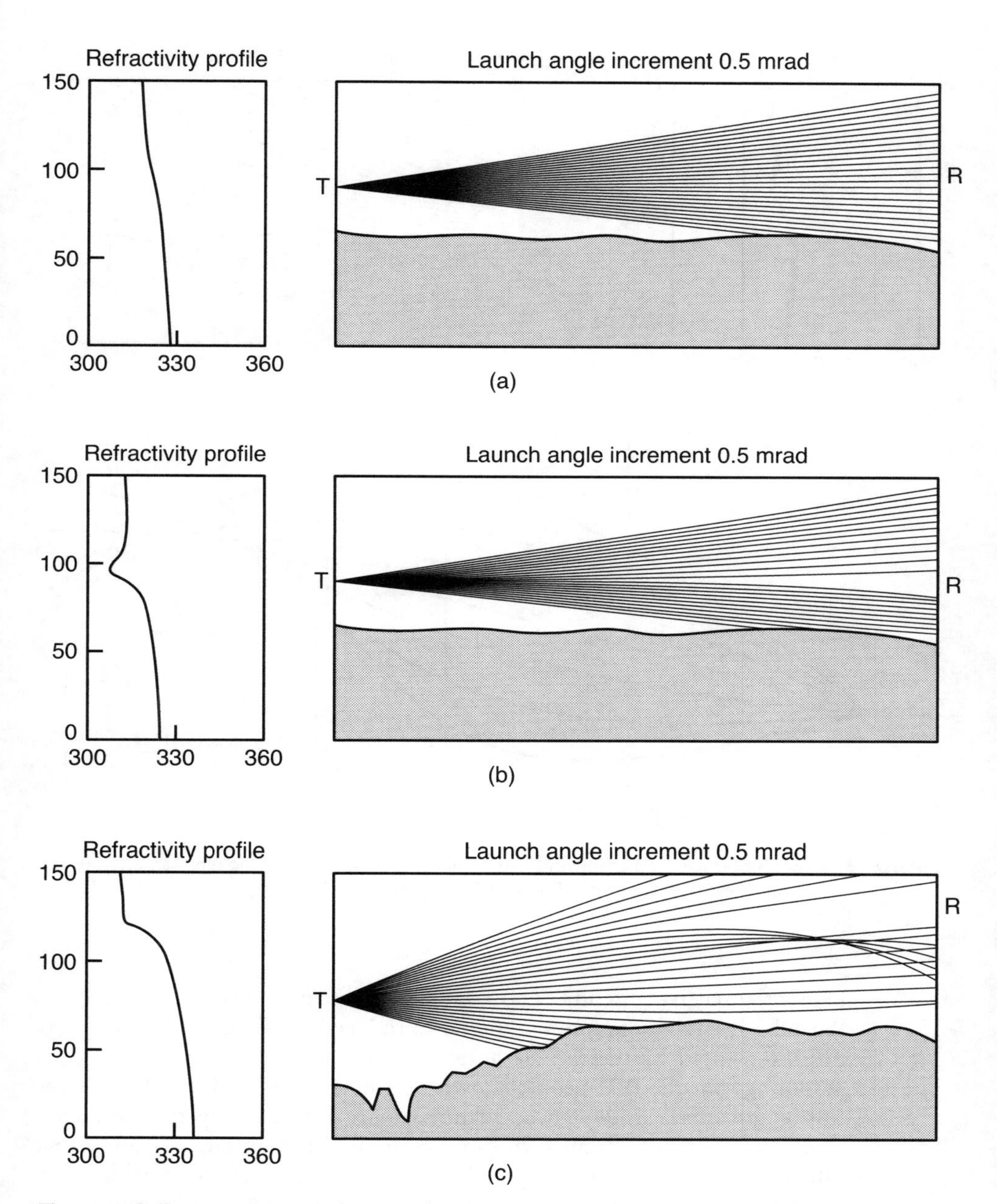

FIGURE 2.5 Ray-tracing diagrams for a range of refractivity profiles: (a) clear-sky situation, (b) defocusing event, and (c) high-low situation.

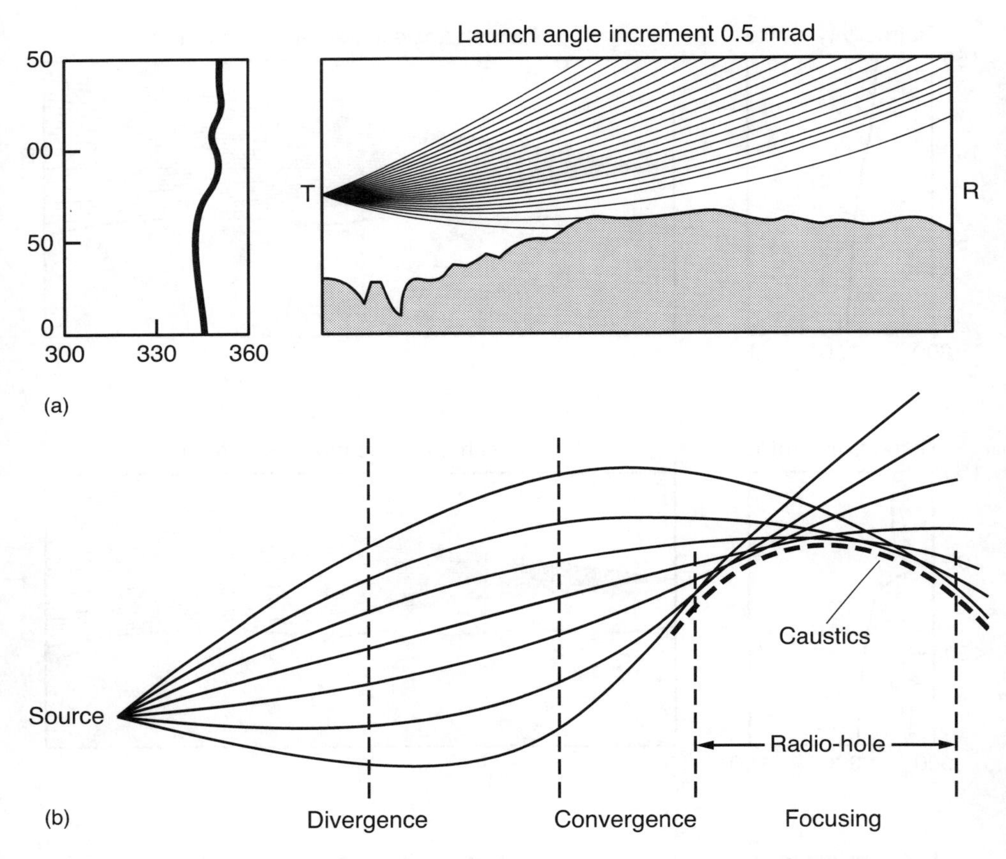

FIGURE 2.6 (a) Ray-tracing diagrams for a subrefractive profile. (b) Graphic representation of certain refractivity concepts. (From CCIR Report 718-3 by kind permission of the ITU.)

below the receive antenna. Those launched above the horizontal pass into the subrefractive region and experience reduced bending and so again are not intercepted by the receive antenna. The ray diagram is produced by assuming that the RRI profile shown on the left-hand side applies for the full length of the radio path. Such a situation is unlikely to apply over more than a short distance in practice, but in the absence of any other information concerning the profile, no other approach is possible. This indicates the difficulties in the ray-tracing approach to determine the cause of observed fading and the uncertainty in propagation experiments.

(b) Figure 2.5(c) depicts the situation in which transmit and receive antennas are at different heights and a severe perturbation of the RRI profile in the height interval between them blocks the signal. This is often referred to as a *high-low situation*. See also section (d) on shadow-zones below.

(c) Figure 2.6(a) demonstrates the losses from ground-based obstruction of the path during severe subrefractive events that can also occur. This situation is avoidable, however, in all but the most extreme situations by using good planning. The planning rules used to avoid this are discussed in Chapter 7.

(d) *Shadow-zones* arise from complex RRI structures, and Figure 2.5(c) represents this situation. The division between a defocus event and a shadow event can be somewhat blurred, depending on the severity of the RRI perturbation, path length, and so forth. Figure 2.6(b), found in CCIR Report 718–3, shows graphically some of the terms used in Chapter 2, although some of the names used in the diagram are common alternatives. Below a certain height in the fourth section from the left, there is no radio energy present—a *radio-hole*. The diagram is idealized, however, and there is residual energy in this area due to scatter from the ground.

2.3 THE EFFECT OF FADING ON SYSTEM PERFORMANCE

2.3.1 Analog Systems

Analog high-capacity systems are relatively narrowband and are not very sensitive to distortion effects caused by multipath propagation. Thus the only factor contributing to degradation of the system performance is the depth of fading. The robustness of any analog equipment against fading is defined by its flat-fade margin. This parameter can be measured by setting up a back-to-back test of the transmitter and the receiver, such that the attenuation between the two results in the receiver input level being the nonfaded or clear-sky value. Additional path loss can then be inserted to represent a fading situation until the signal-to-noise ratio is degraded to the point where the system is no longer usable. The additional loss inserted before this point is reached is known as the *system flat-fade margin*. In practice, the degradation of the system under fading conditions is indicated by a decreasing signal-to-noise ratio, occurring as the system AGC functions and brings up the front-end noise contribution, over the full range of the flat-fade margin. We then describe the system as having a *graceful degradation characteristic*.

2.3.2 Digital Systems

We can evaluate the flat-fade margin of a digital system in exactly the same way except that the bit-error ratio defines the system performance. When this exceeds a value of 10^{-3}, the system is no longer considered usable. A digital system, however, does not degrade gracefully in such a test, with all the degradation occurring as the last few decibels of fading before the flat-fade margin is reached (since the system is very insensitive to flat-fading).

Frequency-selective fading is the critical factor determining the performance of a digital system, with flat fading doing little more than increasing system susceptibility to the multipath problem. The difference between the response of the two systems lies in the bandwidths involved: early high-capacity digital systems operating with channel bandwidths of around 140 MHz compared with analog bandwidths of a few megahertz.

If a narrowband system is operating under multipath transmission conditions, then the system bandwidth is small relative to the spacing between the multipath notches since the delays are only in the order of a few nanoseconds. Thus there is very little variation of amplitude or phase across the channel under consideration. We can consider it as suffering something very close to flat fading.

If the system is wideband, however, there is a significant change of resultant phase across the channel width, resulting in distortion. Figure 2.7(a) depicts such a case in which there is a vector representation of a two-path situation, with the direct path of unity amplitude and the delayed path of amplitude a, where $a < 1$. Since the representation is of a situation with a fixed relative delay, the phase shift between direct and indirect paths is a function of frequency. Hence, the locus of the resultant vector can be scaled in frequency.

Consider the range of frequencies F_1 to F_5. At F_1, the relative delay is such that vector addition occurs. As we move on through F_2 to F_3, the resultant signal level reduces and the phase shift becomes a maximum in the clockwise direction. We then move on through F_4, the frequency at which the minimum signal occurs, to F_5, at which point the phase is at a maximum in the opposite direction. Now group delay is proportional to rate of change of phase, and between F_1 and F_3 there is a gradual increase in phase shift through F_2 and on to the maximum at F_3, followed by a rapid reversal through F_4 to F_5. The group delay will therefore have peaked at the notch frequency F_4. The signal level and the group delay are plotted in Figure 2.7(b).

Figures 2.8(a) and (b) represent a case in which the delayed component is larger than the direct one. Again, the resultant phase shift increases slowly between F_1 and F_3, but after this it continues to increase rapidly up to F_4 followed by a rapid decrease. In this situation, then, the group delay again peaks at F_4 but in the opposite sense.

The first example is known as *minimum-phase fading,* the resultant phase shift being within the limits of $\pm\varphi$. The second example is that of nonminimum-phase fading. Early digital systems with adaptive demodulators responded very differently to the two events, but modern systems do not suffer from this sensitivity.

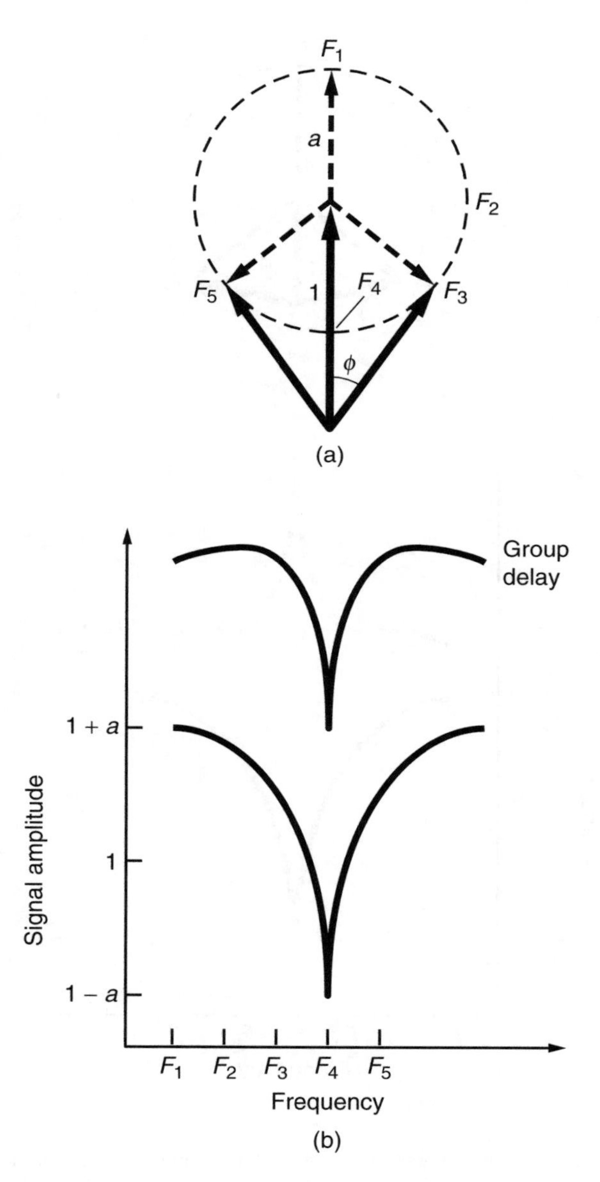

FIGURE 2.7 Mechanism of multipath distortion (minimum phase case): (a) vector representation of a two-path situation and (b) resultant group delay and amplitude characteristics.

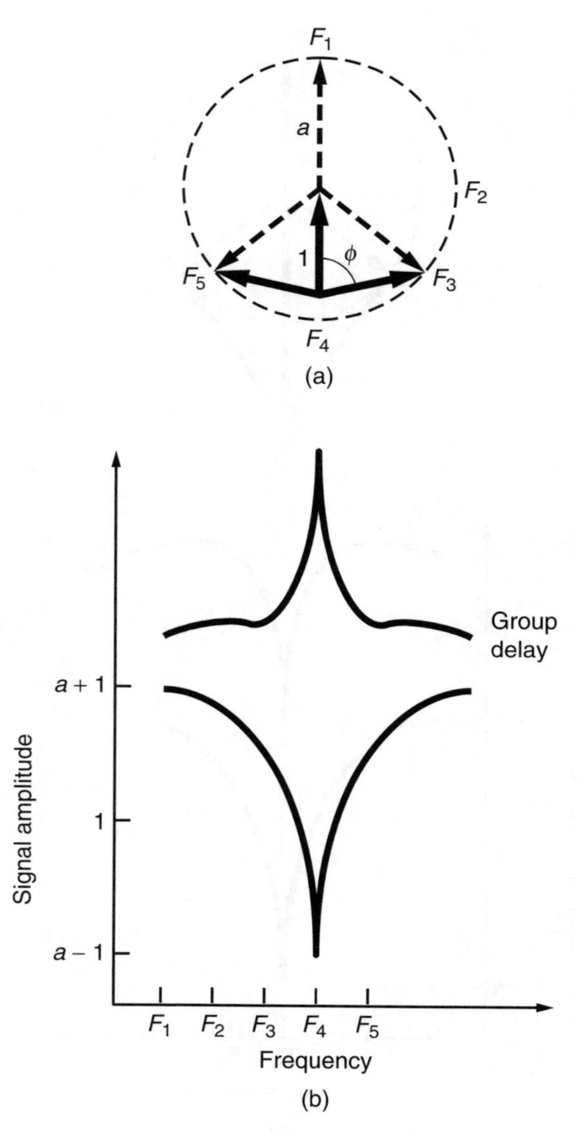

FIGURE 2.8 Mechanism of multipath distortion (nonminimum phase case): (a) vector representation of a two-path situation and (b) resultant group delay and amplitude characteristics.

Two important facts emerge from this exercise. The first is that since the signal level is not constant across the channel for wideband systems, measurement of the AGC voltage does not reveal the true fading situation. Within a narrow bandwidth, however, AGC gives a substantially reliable measure of channel fading. The second fact is that group-delay distortion (also known as *multipath distortion*) is serious if the amplitude of the delayed signal is close to that of the direct one. This was a problem in wideband digital systems, but more robust narrower band systems have since superseded them.

2.4 THE EFFECT OF FADING ON ANTENNA PERFORMANCE

To understand the importance of certain facets of antenna performance, we need to examine the development of frequency plans for microwave systems over a number of years.

The early years of microwave transmission saw a relatively uncrowded spectrum. Those installing systems took advantage of this to reduce system costs. Thus early frequency plans were of the type shown in Figure 2.9(a), in which the RF channels were well-spaced to reduce the complexity of channel filters while preserving the required adjacent channel rejection to avoid interference problems. In addition, the channels were all launched with the same polarization.

As more and more services were introduced into this part of the radio spectrum, the need for better spectrum management was recognized. The first moves in this direction were to close up the channel spacing—made possible by using better channel filters—and then to transmit adjacent channels on alternate polarizations. This latter move meant that the cross-polar discrimination (XPD) of the antennas (defined as the ratio of the signal transmitted and received in a given polarization, to that part of the signal translated into the alternate polarization) could be used to supplement the effect of the channel filters. In Figure 2.9(b), for instance, the adjacent channel interference from channel 2 into channel 1 is reduced first by the rejection offered by the channel filter and second by the antenna XPD. The antenna XPD is not a fixed quantity, however, but is likely to degrade by 1 dB for every decibel of frequency-selective fading present. Thus we cannot rely on the measured XPD under ideal, nonfaded conditions to protect against adjacent channel problems when the system is suffering from multipath fading.

For shallow multipath fades, the cross-polar signal is quite well-correlated with the copolar signal. As the fades become deeper, however, the correlation is less well-marked, resulting in some loss of XPD. The loss of XPD under multipath fading conditions can result from several mechanisms. These mechanisms can be divided into those in which the antenna cross-polar patterns do not play a role and those in which they do. In the first group, a scattering or reflecting mechanism produces a cross-polarized scattered wave in addition to the copolarized wave. In the second group, we

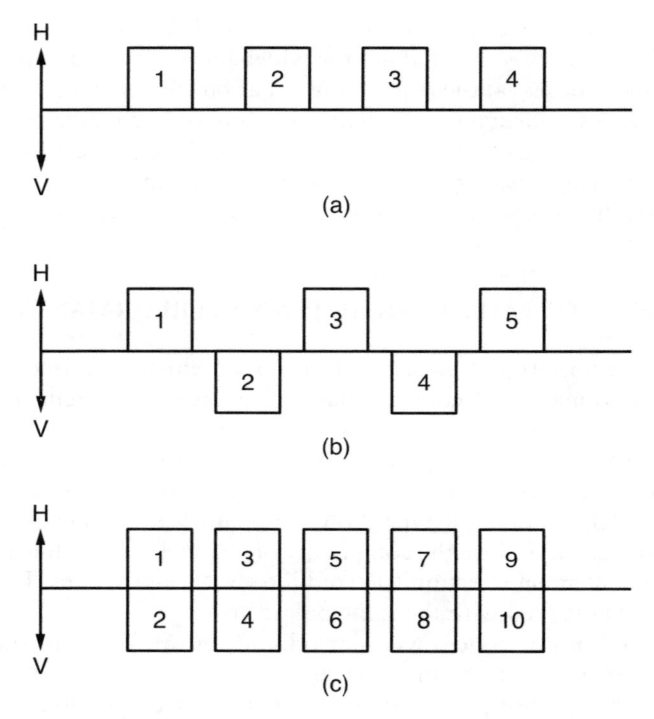

FIGURE 2.9 Development of spectrally efficient frequency plans: (a) wide channel spacing, single polarization; (b) channels closed up and alternate polarization; and (c) cofrequency cross polar.

need an understanding of the copolar and cross-polar patterns of the antenna and how they interact with the multipath components of the received signal.

A simple explanation has been put forward for the two-path situation [3] as depicted in Figure 2.10 in which the signals, of almost equal strength, arriving at the antenna are indicated by S_1 the indirect path and S_2 the direct path. The two components arrive near the peak of the copolar pattern of the antenna and result in a 2-dB difference in signal strength. Hence if they are in antiphase, a deep copolar fade can arise. The same two signals, however, when coupled to the sharp cross-polar characteristic of the antenna, are not equal, and as shown in the diagram have a 15-dB difference. Therefore a corresponding deep fade cannot occur, and the XPD of the antenna is degraded under these conditions.

The final step toward spectrum efficiency is shown in Figure 2.9(c), in which the channels are configured in a cofrequency, crosspolarized mode. In such an

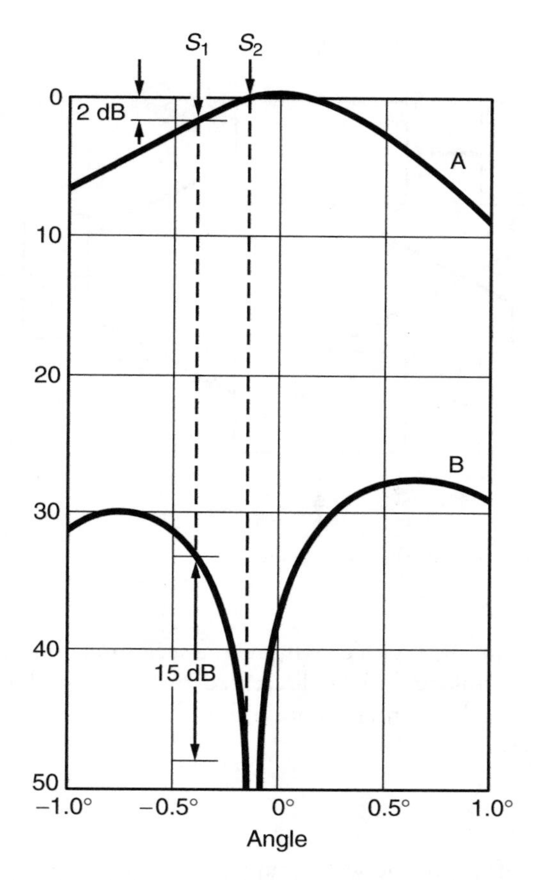

FIGURE 2.10 Explanation of XPD degradation during multipath fading. (S_1, S_2: incoming rays; A: copolar amplitude response; B: cross-polar amplitude response.)
(*Source:* CCIR Report 784–3 by ITU permission.)

arrangement, cofrequency channels rely entirely on the antenna XPD to provide protection against cofrequency interference. High-grade antennas designed specifically for such a situation can offer a nonstressed XPD of 45 dB; thus if a channel requires cofrequency interference to be 19 dB below the wanted signal, such a system should function under selective fade depths of 26 dB. However, observations on an experimental British Telecom link revealed that under fading introduced by some subsidence events, the XPD degraded in sympathy with both frequency-selective and mean-depression components of fading [4]. Cofrequency, cross-polar working is now considered suitable only for systems operating over short distances with relatively moderate

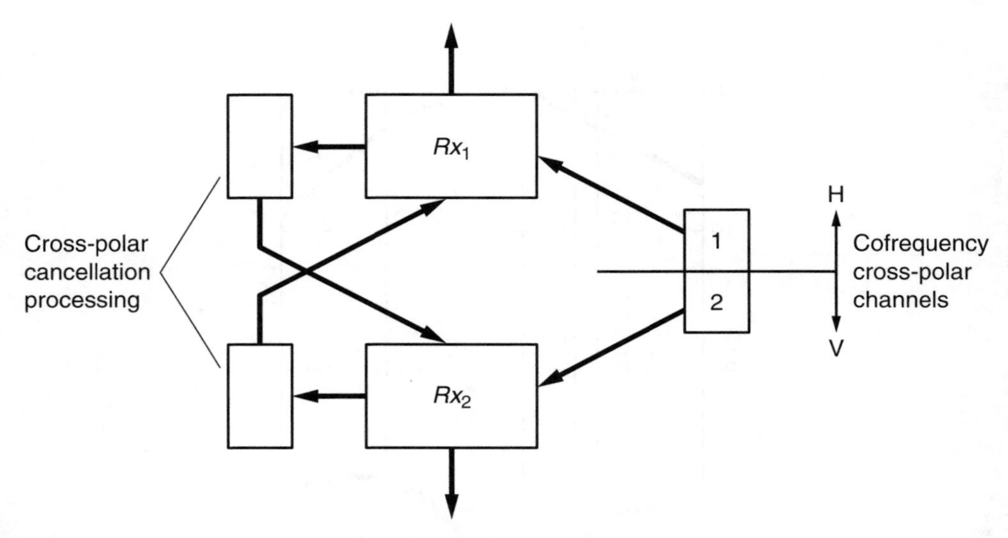

FIGURE 2.11 Cross-polar cancellation technique.

fading, or for those equipped with cross-polar cancellation equipment. In such a system (Figure 2.11), a sample of the signal received on the crosspolarized channel is processed and used to cancel any interference into the cofrequency channel as a result of degraded XPD values.

2.5 SUMMARY

The presence of atmospheric layering can cause fading on microwave links. Study of the fading patterns reveals that there are two components present, a slowly varying component known as the mean-depression term, which is not frequency conscious, and a much more rapidly changing, superimposed component referred to as the multipath term. The latter term is, by reason of the mechanics of its formation, frequency selective. The combined fading pattern is, by general usage, often referred to as being a multipath event, but in truth only the second term is multipath by definition.

Multipath fading is caused by the existence of more than one path by which the signal can travel from transmitter to receiver as a result of the atmospheric layering. In the vast majority of cases this is a two-path phenomenon, whereas mean-depression fading is brought about by bulk refractivity effects. The two components are statistically independent of each another.

Analog systems are found to be virtually insensitive to the type of fading present, the existence of fading being revealed by a decreasing signal-to-noise ratio as the action of the AGC system brings up the front-end noise contribution. Digital systems are, however, virtually immune to mean-depression effects, much of the degradation being brought about by phase distortion as a result of their relatively wideband characteristics.

Antenna XPD, which is used in spectrally efficient frequency plans, is degraded by the presence of frequency-selective fading. This has led to the development of high-XPD antennas and cross-polar cancellation systems.

References

[1] Nomura, et al, "Distortion Due to Fading in High-Capacity FDM-FM Transmission," *Electron Elec Comm Japan,* Vol. 50, No. 6, 1967.

[2] Martin, L, "Relative Amplitude and Delays of Rays During Multipath Fading," *IEE Second International Conference on Antennas and Propagation,* York, United Kingdom, Heslington, April 1981.

[3] Morita, K., "Fluctuation of Cross-Polar Discrimination Ratio Due to Fading," *Rev Elec Comm Labs,* Vol. 19, No. 5–6, pp. 649–652.

[4] Doble, J., "6-GHz Propagation Measurements Over a 51-km Path in the United Kingdom," *AGARD Conference on Terrestrial Propagation Characteristics in Modern Systems of Communications, Surveillance, Guidance, and Control,* Ottawa, October 1986, AGARD CP-407, pp. 20-1 to 20-8.

CHAPTER 3

The Use of Diversity to Reduce System Performance Degradation

3.1 INTRODUCTION

We have shown how, as a result of certain meteorological conditions, a microwave channel can suffer fading that is a combination of flat and frequency-selective components. This fading can be both deep and prolonged. We have also found that the presence of deep frequency-selective fading can degrade antenna cross-polar performance, which in turn may lead to increased cochannel and adjacent channel interference problems. What measures can we take to reduce these problems?

We know that the shorter the length of a microwave link, the less the severity of the fading. Shortening hop lengths to reduce the severity of multipath fading is a very negative approach, however, so what other options are available?

This chapter describes the use of what is known as *diversity reception* to reduce severe fading on microwave systems. There are three types of diversity, and we discuss the benefits and limitations of each. For the most commonly used type, we explain a simple method of determining the improvement factor, in terms of the reduction in the probability of deep fading brought about by its use.

3.2 FREQUENCY DIVERSITY

Since we know that multipath fading is frequency selective, it follows that we can improve system performance if, whenever a channel was degraded by severe fading, we could transfer the traffic being carried on it to a channel operating on a different frequency. We can see that for the situation represented by Figure 2.3 a channel operating at 1.5 GHz would suffer 15 dB of multipath fading besides any mean-depression fading. The multipath notch is very sharp, however, and small change in the frequency at which the channel is operating results in a large reduction in the multipath fade depth. Such an approach is, of course, not spectrally efficient. For this reason, frequency diversity is not recommended by many of the regulatory authorities.

It is interesting to note that someone watching a television channel carrying a program distributed over high-capacity multicarrier microwave circuits can often observe the effect of frequency diversity. In Figure 2.9(b), there are five channels operating as a 4 + 1 system, meaning that there are four traffic-carrying channels and one standby channel (known as the protection channel) to take the traffic from any channel that suffers equipment failure. Now, if a working channel experiences deep fading, the protection switch may, because of the degraded signal-to-noise ratio, interpret this as a system failure and switch the traffic to the protection channel. Thus the program being watched progressively suffers an increasingly noisy picture that will suddenly return to normal as the protection switch operates. This, however, is not the true purpose of the protection channel, and once the switch has operated as a result of a genuine system failure, the protection channel is no longer available to a fading channel.

3.3 SPACE DIVERSITY

A microwave link operating in a two-path environment results in a regular interference pattern up the antenna support structure resulting from the different geometry of the direct and delayed paths at various points on the structure. Figure 3.1(a) is a simplified representation of the situation in which normal ray curvature is ignored and specular reflection from an atmospheric layer is assumed. As the layer's height changes with time, the position and spacing of the interference pattern nulls alters. With a reasonable physical separation of two antennas mounted on the structure, the fading patterns at the antennas can then be considered *decorrelated*. In practice, a separation of 150 to 200 wavelengths—Figure 3.1(b)—can provide good diversity improvement. The actual location of the antennas is discussed under system planning in Chapter 7. It is worth mentioning that the antennas are usually spaced vertically on a single support structure, but they can

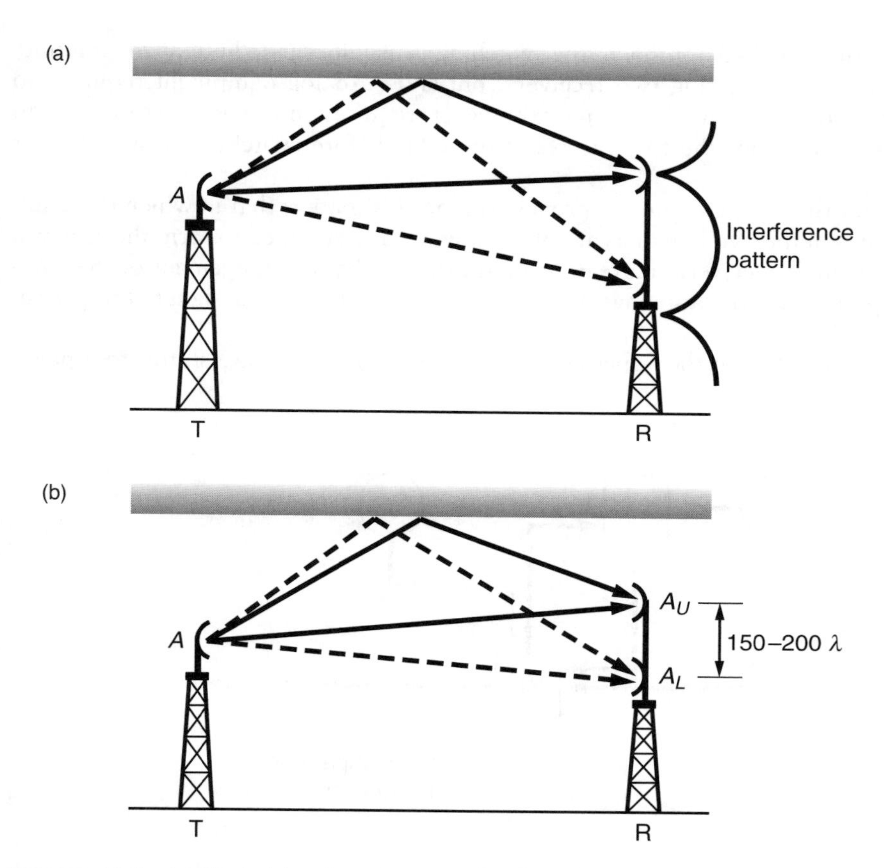

FIGURE 3.1 (a) Vertical interference pattern during multipath fading and (b) required spacing of main and diversity antennas for good diversity action.

equally well be spaced horizontally if there is a need to meet special planning requirements. Whichever orientation is used, the diversity improvement offered is the same for a given separation.

With two antennas with decorrelated fading patterns, how do we use them? There seems to be two possible approaches: either to devise a method by which the receiver is always switched to the antenna with the higher signal level, or alternatively to achieve optimal combination of the two antenna outputs. Regardless of approach, first equalize the lengths of the feeders to the two antennas, adding a balancing length to the lower antenna feed (usually attached to the roof of the equipment room, immediately above the receiver).

For the switched approach, the switching is usually carried out at the intermediate frequency, requiring two receiver front-ends. Analog equipment requires no special precaution. Digital equipment, however, requires an elastic-bit store to align the bit streams from the two receiver front-ends before switching to avoid error bursts.

By far the most common approach to space diversity is that in which the resultant signals on the two antennas are phase-aligned and summed to form the required input to a single receiver. Figure 3.2(a) indicates the basic arrangement of the component parts, with the antennas at the receive end of the link designated upper (u) and lower (l).

The output from the upper antenna passes through a phase shifter to a phase

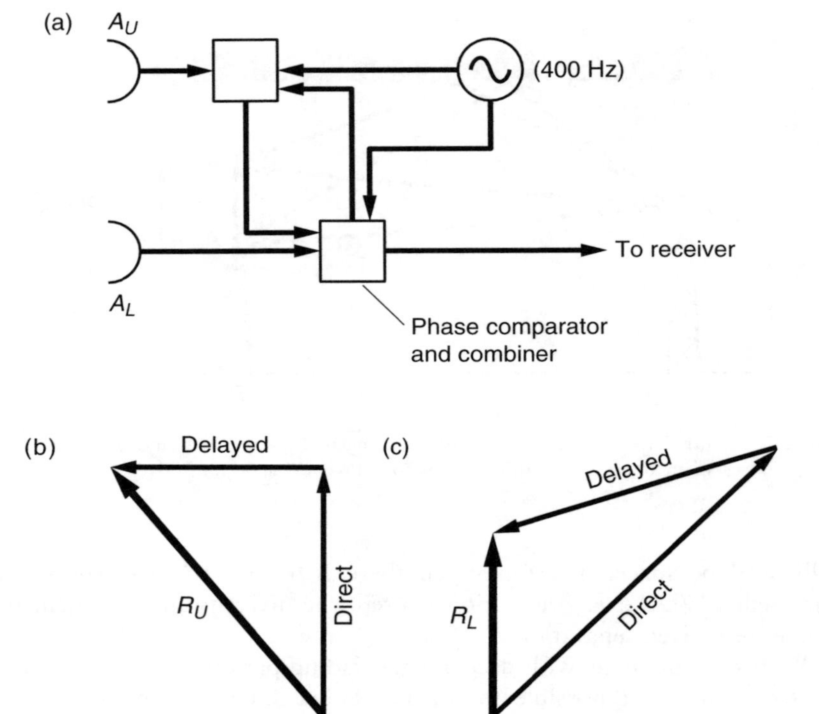

FIGURE 3.2 (a) Signal-combining arrangement for a space-diversity system and (b,c) typical phase relationship of received signal components on the two antennas ((b) upper and (c) lower) of a space-diversity system.

comparator and combiner, a second input to which is the output of the lower antenna. To explain how the diversity system works, let the phase and amplitude of the various signal components be as indicated in Figures 3.2(b) and (c). The upper antenna has direct and delayed components of amplitude and phase as shown (b), and the lower antenna components are as shown in (c). In Figure 3.3(a) the resultants *R*u and *R*l from the two antennas are combined, and *R*u has a phase wobble superimposed upon it. This phase wobble introduces an amplitude modulation of the vector sum of *R*u and *R*l, as the phase of *R*u is driven closer to or farther from that of *R*l. The phase of the amplitude modulation relative to the signal causing the phase wobble allows the direction in which *R*u needs to be driven to phase-align the two antenna outputs to be determined, as follows. Let us suppose that a positive-going signal from the 400-Hz oscillator causes *R*u to move in a counterclockwise direction. This movement increases the phase misalignment between *R*u and *R*l and causes the vector sum of these two components to decrease (i.e., a negative-going part of the amplitude modulation). The relationship between the drive signal and the amplitude modulation is shown in Figure 3.3(b) (in an antiphase situation). However, if *R*u had been, as shown, dotted, the positive-going drive signal would then have driven *R*u and *R*l nearer to phase alignment, and the amplitude modulation would have been in phase with the drive signal (Fig. 3.3(c)). Hence these phase relationships can determine the sense in which the phase shifter in the output from the upper antenna needs to be moved to phase-align *R*u and *R*l. (Once the two vectors are aligned, there will be no 400-Hz component in the amplitude modulation, since whichever way the phase of *R*u moves relative to *R*l causes a reduction in the vector

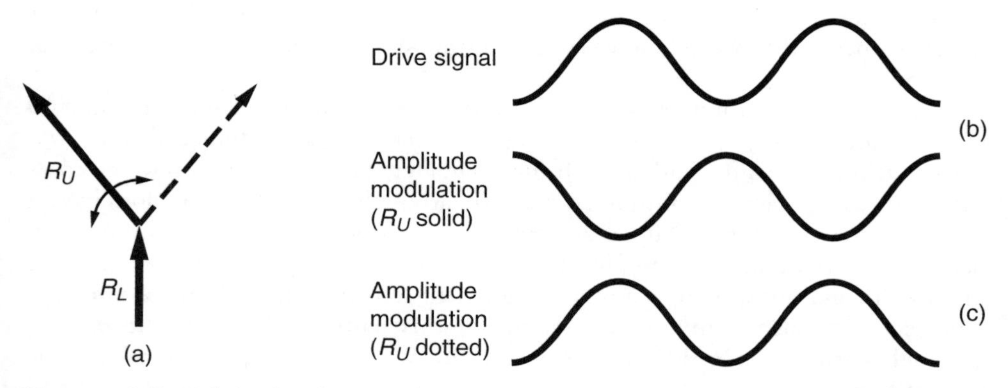

FIGURE 3.3 Relationships between phase and amplitude modulations in a space-diversity system: (a) summation of signals from upper and lower antennas, (b) phase relationship between drive signal and resultant amplitude modulation (R_U solid), and (c) phase of amplitude modulation (R_U dotted).

sum of the components and changes the amplitude modulation under this condition to an 800-Hz frequency.) All the necessary information is available, then, to enable continuous optimization of the phase alignment of signals on the two antennas of the diversity system.

Understand that the purpose of installing diversity systems is to minimize the multipath fading experienced by the receiver. Any consequential reduction of the flat-fading component is fortuitous. The reduction of multipath not only improves the signal-to-noise or the bit-error ratio of the system but also reduces the degradation of the XPD, leading to improvement in the areas of cofrequency and adjacent-channel interference. Figure 3.4 is an excellent example of the reduction in multipath fading caused by using space diversity.

3.3.1 The Effect of Space Diversity on the Severity of Narrowband Fading

When planning an analog microwave system, it is necessary to have information on the likelihood of multipath fading exceeding certain levels to ensure that the system can achieve its performance targets. This information is usually presented graphically in the form of a cumulative distribution of deep fading: defined as the fading exceeded for less than 0.1% of the worst month of an average year. The level of fading exceeded for 0.1% of the worst month is known as the $F_{0.1}$ value. A model for estimating it is in Chapter 6. This raises the question, however—what is meant by the *worst month*?

To answer this, let us say that someone has records of nondiversity fading on a particular link and they plot the cumulative distributions of fading for each month of the year on the same set of axes. Then the worst-month characteristic is the envelope of the twelve curves, with a mean slope of around 10-dB/decade of probability.

To avoid performing this exercise every time you are involved in planning a link, the nondiversity worst-month fading characteristic is a 10-dB/decade line originating at the $F_{0.1}$ value. Fading with this characteristic is sometimes referred to as *Rayleigh fading*. This is a misnomer, because although Rayleigh fading does have a 10-dB/decade slope, it is defined as being the vector sum of a number of components of random phase and amplitude. In microwave LOS systems operating under multipath conditions, there is a main component (direct path) and one or two randomized components (multipath). Fading of this nature is, in fact, *Rician*, and although it still has a 10-dB/decade slope, it differs from Rayleigh in the tails of the distribution.

There are a number of methods to determine the improvement factor from using space diversity. These tend to deal with the less common situation in which the two antennas used are of different sizes, however. An empirical approach to

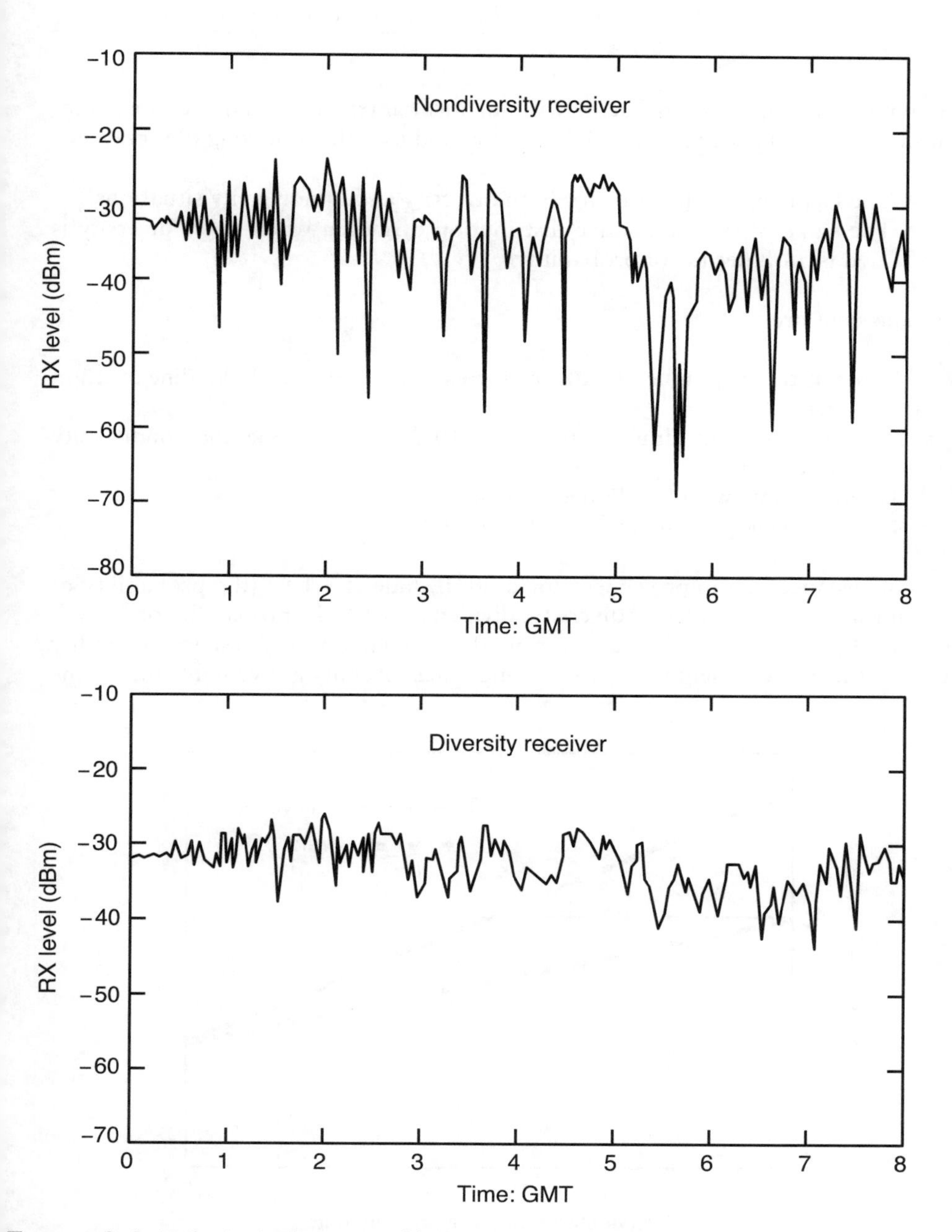

FIGURE 3.4 Reduction in the severity of fading by the use of space diversity.

determining the improvement factor with identical antennas (by far the most common situation) is shown in Figure 3.5. The method uses the following observations:

- The $F_{0.1}$ figure is the same for both diversity and nondiversity situations.
- The slope of the cumulative distribution of fading when space diversity is used is 5-dB/decade of probability.

The steps are:

1. Mark the $F_{0.1}$ value (15 dB in this example) on 3 cycle log/linear graph paper.
2. From this point, draw a line with a 10-dB/decade slope, the nondiversity characteristic.
3. Draw a line with a 5-dB/decade slope.
4. Read the improvement factor horizontally.

For instance, the improvement for a 30-dB fade is 3.10^{-3} (the probability of exceeding a 30-dB fade without diversity) divided by 10^{-4} (the probability of exceeding a 30-dB fade with diversity), a factor of 30. For a narrowband system, this ability to predict the improvement factor from using space diversity is a valuable tool in the

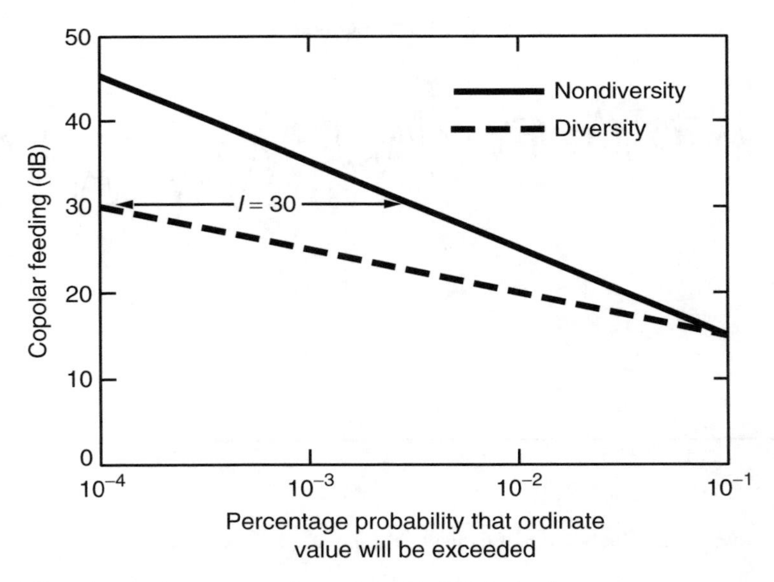

FIGURE 3.5 Diversity improvement factor I at a 30-dB fade depth.

prediction of system outage (as is shown in Chapter 7), since outage in narrowband analog systems is directly related to the probability of fade exceeding the system flat-fade margin. Thus if the target-system-outage probability is less than that of the fade exceeding the system fade margin in a nondiversity situation, clearly diversity is required. You can easily determine the improvement brought about by diversity.

Few techniques exist for estimating the diversity improvement factor for wide-band digital systems. One technique [1] is based on a simplified three-ray model. Another, [2], is based on the relationship between statistics of linear amplitude dispersion (LAD) for diversity and nondiversity situations. We deal with this area more fully in Chapter 6.

3.4 ANGLE DIVERSITY

We should consider two techniques under this classification. The first of these is the use of antennas with two or more beams separated by small angles in the vertical plane [3] (or, alternatively, separate antennas angularly offset in the vertical plane). Because of the different angles of arrival, the multipath components add up in a different way for the different beams, consequently resulting in different frequency-selective fading in the diversity branches. Some diversity improvement has also been observed for the flat-fading component [4].

The second method states that in an angular separation between the direct and delayed signal paths, it follows that fading could be reduced somewhat if the antenna boresight were always angularly aligned with the direct signal path, whatever the angular offset from normal due to bulk-refractivity changes (Fig. 3.6(a)). An

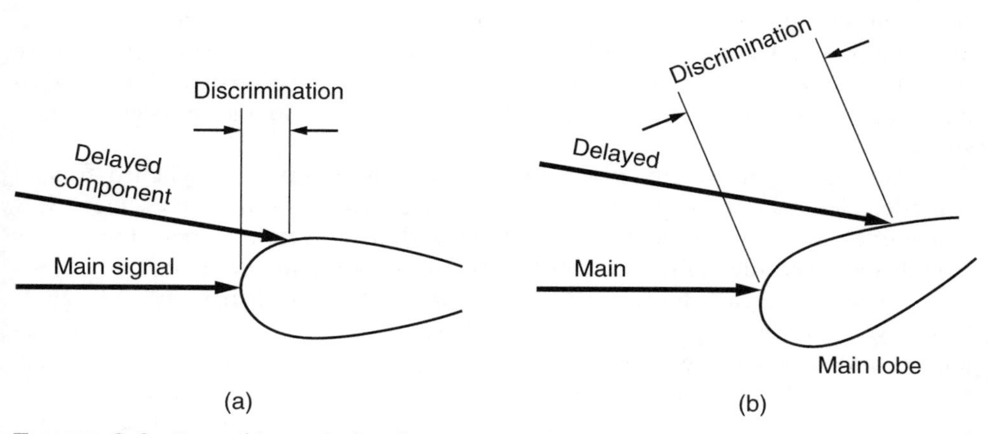

FIGURE 3.6 Beam-tilting angle diversity.

extension to this would move the boresight so that it lies to the side of the direct path remote from the delayed component, such that although the main signal suffers some attenuation, the delayed path suffers much more because of the steep sides of the main lobe pattern of the antenna. This then reduces the degree of cancellation (Fig. 3.6(b)). However, until electronic adaptively steerable antennas are fully developed, this type of diversity is unlikely to find wide use.

3.5 RELATIVE MERITS OF THE THREE TYPES OF DIVERSITY

One main advantage of space and angle diversity over frequency diversity is that they are spectrally efficient.

The most obvious factor in favor of single-antenna angle diversity is reduced cost and tower loading. Also, in terms of relative performance on digital links, angle diversity may give the greatest improvement on links in which distortion-induced outage dominates. However, space diversity may give the greatest improvement on links in which flat-fading–induced outage dominates [4–7]. In narrowband systems, space diversity gives the greatest overall improvement.

For the problem of strong, sea-reflected multipath, there is evidence that diversity reception between a nontilted beam and a beam tilted toward the sea-reflected wave is effective [8,9].

3.6 SUMMARY

Severe multipath fading can lead to system outage, degraded antenna cross-polar performance, and adjacent channel and cochannel interference. Although a reduction in hop length reduces the severity of these problems, such a move is a very negative approach. A good engineering approach uses the benefits offered by diversity reception, using three variants: frequency, space, and angle diversity. All these variants aim to reduce multipath fading, as well as flat fading, to a lesser degree, under some conditions.

The first of these, frequency diversity, in which traffic is switched from a degraded channel to one operating at an offset frequency, is simple to engineer (it is a side effect of operating multichannel links with a protection system) but suffers from being spectrally inefficient. Hence regulatory bodies do not favor it.

Space (or height) diversity is a proven method in which the received signals from two antennas, separated by 150–200 wavelengths, are continuously phase-aligned and summed. We know that the fading patterns on antennas at this spacing are decorrelated, so the summing process ensures that a near optimal signal level is presented to the receiver for nearly 100% of the time. A variation on this, in which one antenna is angled downward to point at a ground reflection, has effectively reduced problems encountered when crossing water, marshland, or the like.

Angle diversity, comparatively still in its infancy, uses either the different fading patterns received by antenna feeds angularly offset from each other, or the reduction of the delayed components by tilting the receive antenna and artificially increasing the discriminatory effect of the main lobe pattern.

References

[1] Rummler, W. D., "A Statistical Model of Multipath Fading on a Space-Diversity Radio Channel," *BSTJ,* vol. 61, 1982, pp. 2185–2219.

[2] Higuti, L., and Morita, K., "Diversity Effects of Propagation Characteristics During Multipath Fading in Microwave Links," *Rev Elec Comm Labs,* vol. 30, no. 3, 1982, pp. 544–551.

[3] Dombek, K. P., "Reduction of Multipath Interference by Adaptive Beam Orientation," *Proc European Conference on Radio-Relay Systems,* Munich, November 1986, pp. 400–406.

[4] Valentin, R., et al, "Space Versus Angle Diversity—Results of System Analysis Using Propagation Data," *ICC 1989, IEEE.*

[5] Lin, E.H., et al, "Angle Diversity on Line-of-Sight Microwave Paths Using Dual-Beam Dish Antennas," *Proc. Int. Conference on Communications,* Seattle, 1988, Paper 23.5.

[6] Alley, G. D., et al, "The Effect on Error Performance of Angle Diversity in a High-Capacity Digital Microwave Radio System," *IEEE Globecom 1987.*

[7] Mohamed, S. A., et al, "Results of Angle Diversity Trial by British Telecomm," *Second European Conference on Radio-Relay Systems,* 1989.

[8] Sasaki, O., et al, "A Tilted-Beam Diversity Reception System to Reduce Line-of-Sight Microwave Fading," *Trans Inst Electron Comm Engrs, Japan,* 70-b, 10, 1987, pp. 1251–1253.

[9] Satoh, A., et al, "Improvement of In-Band Amplitude Dispersion by Beam Tilting on Radio Links With Strong Ground Reflection," *Proc Intern Symp, Antennas, and Propagation,* 1989.

CHAPTER 4

PROPAGATION-INDUCED INTERFERENCE

4.1 INTRODUCTION

Although one can design a microwave radio system to have the required degree of interference immunity under clear-sky conditions, the effects of some propagation phenomena can considerably erode safety margins. A system designer must have a full understanding of the mechanisms by which this erosion occurs to achieve a successful system performance under adverse propagation conditions.

This chapter discusses three types of interference, the conditions under which they may become a problem, and possible countermeasures. The three types of interference are:

Adjacent channel interference (ACI) is interference from the small frequency separation of other channels in the spectrally efficient frequency plans used in high-capacity microwave systems.

Cochannel interference (CCI) arises either from reuse of the channel frequency within the radio network or from working in a cofrequency, cross-polar mode with insufficient margins.

Nodal-point interference (NPI) results from a number of routes with uncorrelated fading patterns and small angular separations converging on a single location.

4.2 ADJACENT CHANNEL INTERFERENCE

Using channel filters offers protection against ACI, with additional protection from antenna XPD in the case of interleaved channel plans.

Looking first at the channel filter protection in Figure 4.1, in which the filters are represented by the idealized characteristics, shows that the ACI arises in the crossover region of the filters. The sharpness of the filters used, and hence the cost, is chosen to offer protection under clear-sky conditions, with a suitable safety margin for multipath situations.

Now if a channel is stressed by a multipath notch within its bandwidth, you can then see that the adjacent channels are faded to a lesser degree, due to the frequency-selective nature of multipath fading. Hence the relatively higher level of the adjacent channels results in a higher level of ACI in the stressed channel, and the designer must account for this degradation from multipath activity when determining the safety margins.

When an interleaved channel plan is used, the relative level of the adjacent channels is reduced by the antenna XPD. Although it is possible to use this to reduce the sharpness of the channel filters, the need for spectral efficiency means that the chance to make a reduction in channel spacing takes priority, thus negating the whole benefit offered by the XPD to reduce ACI. Also, as explained in Sections 1.2 and 2.4, the clear-sky XPD figure is degraded on a decibel-for-decibel basis with the frequency-selective component, so this XPD erosion increases the level of ACI.

Thus, when determining the filter characteristics the designer must consider the relative enhancements of the adjacent channels, brought about by the existence of

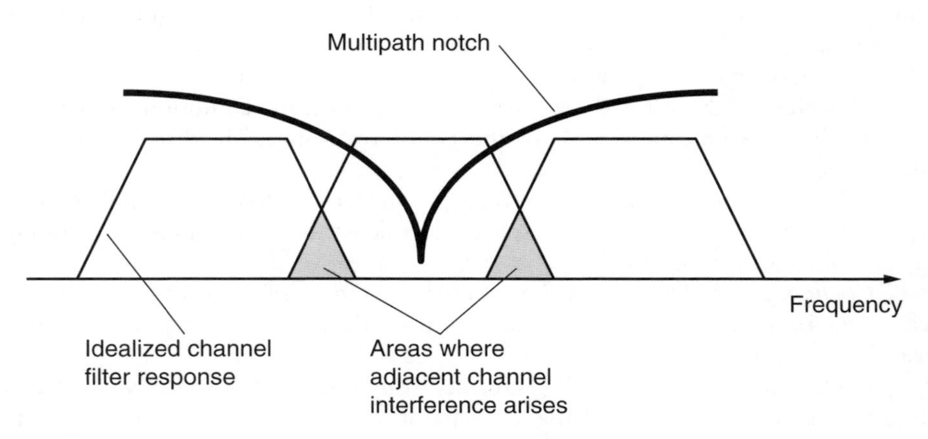

FIGURE 4.1 ACI in the crossover region of the channel filters.

multipath activity, keeping in mind the possible degradation of XPD as a result of this activity.

4.3 COCHANNEL INTERFERENCE

A number of scenarios can cause CCI:

- Bad route planning;
- Failure to account for other routes in the same country;
- Failure to carry out international coordination procedures;
- XPD degradation in a cofrequency, cross-polar frequency plan.

4.3.1 Bad Route Planning

A microwave route extending over 200 to 300 km comprises a number of sections around 50 km long. As a countermeasure against interference, the signal polarization is reversed at each repeater, and a dogleg is inserted to avoid overshoot interference from one section of the route affecting a repeater in a later section (a form of CCI). In Figure 4.2(a), a badly planned route results in repeater D being on the overshoot path of section A to B. Although under normal conditions, the signal from A into D suffers a considerable amount of attenuation because of the long path involved, together with a further loss due to the offset arrival angle at D. When ducting occurs the path loss involved becomes much less, with interference experienced at repeater D from the signals arriving by two different paths, both signals having the same polarization. In Figure 4.2(b), the relocation of repeater D results in an additional offset loss on the path A–D, since D no longer falls on the overshoot path.

4.3.2 Failure to Account for Other Routes in the Same Country

When planning a new route, ensure that no repeater is situated on the overshoot path of any section of its own or any other route and that no section of the new route is aligned such that its overshoot path interferes with any repeater on an existing route. Failure to follow these guidelines results in degraded performance from CCI.

4.3.3 Failure to Carry out International Coordination Procedures

Just as one is careful to avoid problems with other routes in the national network, there are requirements to coordinate the introduction of new links on an international basis [1]. The true severity of interference over long distances from ducting situations was not understood until relatively recently. Because the CCIR initiated extensive measurement programs, however, internationally agreed coordination procedures are now in place for compliance by all administrations.

The common factor underlying these three scenarios is the presence of ducting situations. This is not the whole story, however. In 1981, while examining the current methods for estimating the severity of interference, the assumption that no correlation existed between the interference into a terrestrial LOS receiver due to prop-

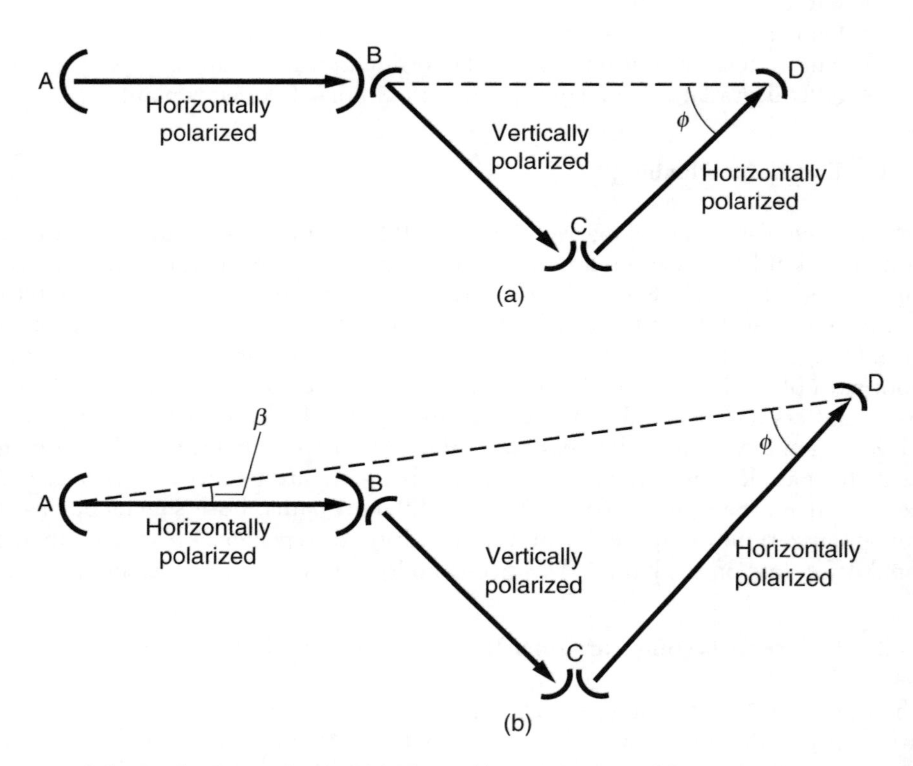

FIGURE 4.2 Doglegging used to avoid CCI caused by overshoot: (a) If the antenna discrimination for the offset angle ϕ, between the overshoot path from A to D and the wanted path from C to D is insufficient, then overshoot interference could be experienced, especially during periods when ducting is present. Note that polarization reversal is used at each repeater. (b) In the example of overshoot interference shown in (a), the poor route planning has been emphasized by placing terminal D directly on the line of shoot between A and B, so that the only overshoot protection is offered by the offset receive angle ϕ of the overshoot path relative to the wanted path. By moving D away from the overshoot path, additional protection is offered by the offset launch angle β between paths A to B and A to D.

agation along the great-circle path from another station transmitting in the same band, and the fading of the wanted signal at the terrestrial receiver was questioned [2,3]. From extensive databases of signal levels on a transhorizon path and those on a LOS link terminating in the same area, it was shown that there was a strong negative correlation between the two signals (i.e., an enhanced transhorizon signal often coexisted often with fading on the LOS path). Thus the signal degradation from CCI could be significantly more severe than originally believed. Canadian measurements in 1991 subsequently supported this view [4,5].

4.3.4 Degradation of XPD in a Cofrequency, Cross-polar Frequency Plan

As we described in Chapter 3, frequency plans using a cofrequency, cross-polar (CFCP) format are viable although sensitive to XPD degradation under multipath fading conditions. High-grade antennas with clear-sky XPDs of around 45 dB are available. Thus if frequency-selective fading of 25 dB is experienced, a carrier-to-interference ratio of around 20 dB is still possible when working in a CFCP format. Anomalous fading events experienced in the United Kingdom, however, [6], suggest that under certain conditions both flat and frequency-selective fading components can cause degradation of XPD so that, for long links at least, cross-polar cancellation techniques are necessary to keep CCI at tolerable levels.

4.4 NODAL-POINT INTERFERENCE

Where several routes terminate at a nodal point, we find that the angular separation between some of them is such that the antennas do not provide sufficient discrimination to avoid the problem of CCI, especially in periods when the wanted signal is deeply faded. This geometry is shown in Figure 4.3, in which the wanted signal path is indicated by the solid line and a possible interferer by the interrupted line. Suppose

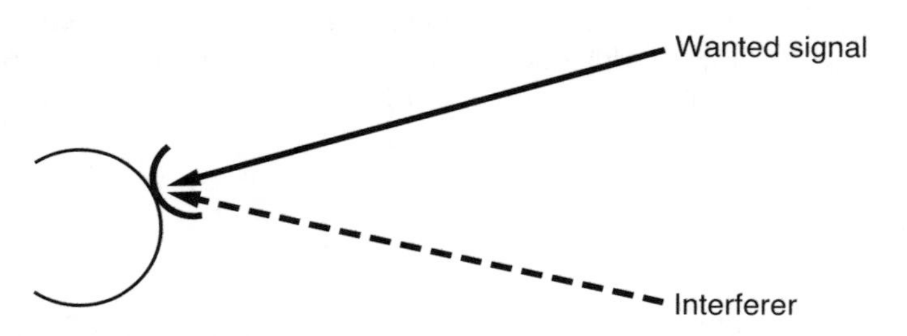

FIGURE 4.3 The geometry of NPI.

that the interfering signal enters the antenna by way of a side lobe that is, say, 40 dB below the main lobe. Now there is no reason why the fading pattern of the two incoming signals should correlate, so if the wanted signal is subjected to a frequency-selective fade of, say, 20 dB, and the interferer is not faded, then the carrier-to-interference ratio is reduced to 20 dB if you assume that the flat fading on the two paths is equal—a commonly used approach to estimate the problem's severity. (The argument for this is that since the two signals are approaching from the same general direction, the effects of atmospheric layering are common to both paths. However, this is a very dangerous assumption, especially in situations with different path slopes.) Thus the 20-dB carrier-to-interference ratio calculated above is a *maximum* value, with even greater degradation if the flat fading on the wanted signal is greater than that on the interferer.

To minimize this problem:

- If possible, arrange for the wanted and interfering signals to be crosspolarized.
- Use a custom-built or adaptive antenna that can provide a sharp null in the direction of the interferer.
- Use a power-control, in which the transmitted level of the wanted signal from the last repeater into the nodal point is continually adjusted in such a way that the received level is always the clear-sky value. (Since the radio path between the repeater and the nodal point is reciprocal, monitoring of the fading on the nodal point to repeater hop allows continuous determination of the correction required to the transmitter power in the reverse direction.)
- Make the section of the route terminating at the nodal point relatively short. We find in Chapter 7 that the UK multipath prediction model, in common with several other models, contains a term 35Log *d* where *d* is the path length in kilometers. Thus the use of a final hop of length 8 km means that the multipath fading predicted for any given percentage time is very much smaller than that predicted on a 50-km hop, reducing the nodal-interference problem. One side effect is that with very low levels of multipath fading, there is no need to use height diversity, thus much reducing the windage loading of the nodal-point antenna support structure. This last approach to reducing the problem of NPI is the most commonly used.

4.5 SUMMARY

Although it is reasonably simple to design and equip a microwave route so that it meets the required interference criteria under clear-sky conditions, the planner must be aware that certain propagation effects can significantly modify the situation. Unless you consider the effects caused by multipath transmission and large-scale ducting events, the system's carrier-to-interference ratio can be seriously degraded.

ACI arises in the crossover region of the channel filters. The filter cost relates directly to the sharpness of the cutoff, so that the choice of filter characteristic is a tradeoff between cost and performance. If, under fading conditions, a channel is stressed by the presence of a multipath notch within the channel bandwidth, then the adjacent channels are somewhat less faded, because of the frequency-selective nature of the fading and contribute significantly more to ACI. If the system is running with an interleaved frequency plan, then under multipath conditions, XPD degradation contributes further to the interference problem.

CCI can be caused by overshoot energy from remote sections of the link on which the problem occurs, being received by a repeater farther down the link. This is usually a case of bad planning. As an alternative, the overshoot energy can arise from another system altogether during periods when a ducting situation is responsible for a long-distance, low-loss interference path, but once again good planning and coordination with other administrations usually avoid this situation.

NPI can arise when energy from an unwanted source enters a system through a side lobe of the antenna. Normally such a signal is several tens of decibels below the wanted signal, but under severe differential fading situations, this can change drastically. There are several ways to reduce the problem, but the most common one is to make all link sections operating into a node, of very short length, so that deep multipath fading is eliminated.

Thus, when designing systems and planning routes, you should be fully aware of the multipath effects you are likely to encounter. Remember, excess path loss is not the only system degradation possible, and it is the differential aspect of the fading, resulting from the frequency-selective nature of multipath activity, that is responsible for some of the untoward events.

Always consider the planning of a new route with the total network in mind. Links that are a great distance apart and appear quite incapable of causing mutual interference problems can, under ducting situations, suffer severe and prolonged problems. You should pay close attention to coordination procedures.

References

[1] CCIR, *Reports of the CCIR 1990,* ITU, Annex to Vol. 5, Geneva 1990, Report 569–4.

[2] Doble, J.E. "Interference on Digital Radio Links due to Trans-Horizon Propagation," *Electronics Letters,* Vol. 17. No. 12, June 1981, pp. 399–400.

[3] Doble, J.E., "Overshoot Interference on Microwave Radio Links due to the Co-Existence of Multipath Fading and Trans-Horizon Propagation," *AGARD Conference on Propagation Aspects of Frequency Sharing, Interference and System Diversity,* Paris, October 1982, pp. 21-1 to 21-7.

[4] Olsen, R. L., et al., "Observed Negative Correlation Between an Interfering Signal on a Trans-Horizon Path and Wanted Signals on Terrestrial Line-of-Sight Paths," *Electronics Letters,* Vol. 27, February 14th, 1991, pp. 332–334.

[5] Bilodeau, C., et al., "Observed Signals on a Trans-Horizon Path and Line-of-Sight Paths: A Case of Negative Correlation," *SBMO International Microwave Symposium,* Rio de Janeiro, July 1991.

[6] Doble, J.E., "6GHz Propagation Measurements Over a 51-km Path in the United Kingdom," *AGARD Conference on Terrestrial Propagation Characteristics in Modern Systems of Communications, Surveillance, Guidance and Control,* Ottawa, October 1986, pp. 20-1–20-8.

CHAPTER 5

RAIN ATTENUATION AND MOLECULAR ABSORPTION

5.1 INTRODUCTION

Fading can occur on LOS paths as a result of absorption and scattering by snow, hail, fog, and rain. For the very small percentage of time that is of interest in system design, however, and for the range of frequencies in general usage, we need only account for rain attenuation normally.

Snow consists of the aggregated ice crystals and large flakes that form just below freezing. Dry snow does not seem to have any marked effect on radio-wave propagation below 30 GHz, partly because of its low-density structure and partly because the flakes tend to tumble without any preferred orientation. Tests show that although wet snowfall can cause larger attenuation, these events do not tend to dominate the attenuation statistics. The degradation of antenna characteristics from snow and ice buildup on the surface of the antennas, forming a modification to the reflecting surface and affecting antenna directivity, may well be of greater importance than snowfall along the path. To avoid such problems, protective radomes cover the antennas in climates where you can expect events of this nature.

Hail forms by supercooled cloud droplets that give rise to large, high-density spherical accretions. They are only weakly attenuating below 30 GHz, however, and they demonstrate little in the way of polarization effects.

Studies show that at frequencies around 100 GHz and above, attenuation in fog may become significant, with the specific attenuation at 140 GHz at around 0.4 dB/km for medium fog and at 4 dB/km for thick fog.

In addition to the above effects, consider attenuation from molecular absorption by oxygen and water vapor that can become significant at some frequencies.

In view of the above comments, we consider only attenuation from rainfall and molecular absorption within this chapter.

5.2 POLARIZATION AND FREQUENCY SENSITIVITY OF RAIN ATTENUATION

Rain attenuation is polarization-conscious from two main causes. First, as a result of the drops not being truly spherical, horizontally polarized waves suffering greater attenuation than those vertically polarized [1,2] second, especially during heavy periods of rainfall, the drops fall with a significant canting angle because of the strong winds usually associated with such events. These two effects can also result in cross polarization components being generated when a LOS signal travels through rainfall that can cause a problem for systems using an interleaved channel plan.

The frequency-conscious nature of rain attenuation, which is related to the drop-size distribution [3,4] is such that you need not consider it at frequencies below 11 GHz for regions having a rain climate similar to that of northern Europe. For monsoon climates and regions suffering from severe storm problems, however, the critical frequency is as low as 5 GHz. Above these critical frequencies, excess attenuation becomes progressively more of a problem, limiting your use of the path length. (It is sometimes stated that above the critical frequency, multipath becomes less of a problem. This is not so, for the multipath-prediction model has a term 8.5 Log F. The misunderstanding arises because the path lengths used at frequencies above critical must be reduced to cater for the rain attenuation. Since multipath is very sensitive to path length, it eventually ceases to become the dominant factor in link performance.) It should be noted however, that multipath fading and rain attenuation are mutually exclusive events (since heavy rain causes mixing within the atmosphere, thereby breaking up any layering present) and therefore outages from the two causes have to be summed when assessing the link performance.

5.3 SOURCES OF RAIN-RATE DATA

The calculation of rain attenuation on a microwave link requires a knowledge of rainfall rates in the range 10^{-1} to 10^{-4} percent of time at the location of interest.

Note that this information is in no way related to annual rainfall figures, so where do we obtain this data?

The first possibility is from the Meteorological Office, which may have straightforward records of this type. As an alternative, you may obtain the information by processing data from any area flood studies. Even if the data are only available for one of the percentages of time, the CCIR has a model that can predict attenuation figures for percentage times in the range 1.0 to 10^{-3} [5].

If reliable information is not available from the meteorological office, then you can have recourse from CCIR data [6]. This can be in the form of rainfall contours for 0.01% of the time, which you can then use in the above model or in the form of rain-climatic zones for which rainfall-intensity figures are given for a number of percentage times. A word of warning, however—the data from this source represent the best available at the time of printing, but are, of course, only as good as the input data on which they are based. As an example, the rain-climatic zone in which Singapore is found includes Malaysia, Indonesia, and part of Northern Australia, and there is no way in which this can be correct. Shortage of input data negates any possibility of breaking this zone down into smaller, more accurate zones.

Finally, there is the possibility of collecting the data with a network of rain gauges. The year-to-year variability of rainfall rates means that this is a slow process. The rain gauges used in such an exercise could be of the tipping-bucket variety, in which a funnel captures the rain and guides it to a pivoted bucket that becomes unstable once it gathers a certain amount of water. The gauge then tips the contents out and records the event on a counter. This device has a typical 2-min resolution. In contrast, a modern drop-counting device has a 10-sec resolution. In this type of gauge, the water collected in the funnel is formed into standard-sized drops that it counts as they interrupt a light beam as they fall.

Remember, predictions using measured data are always more reliable than those obtained by any form of processing information from a single percentage time.

5.4 USING RAIN-RATE DATA TO CALCULATE PATH ATTENUATION

There are several proposed methods for predicting rain-attenuation statistics from rainfall rate measurements made near the path. These methods all use the relationship between the specific attenuation and the rain rate, which depends on the rainfall microstructure. The main difference between the methods lies in the models used to describe the time-space structure of rainfall rate [7].

When predicting rain attenuation, you must appreciate that the intense rain that causes the extremes of attenuation for which the link must be planned occurs in cells of relatively small dimensions (typically less than 5 km). The characteristic model of a rainstorm, often referred to as the *Mexican hat,* is shown in Figure 5.1.

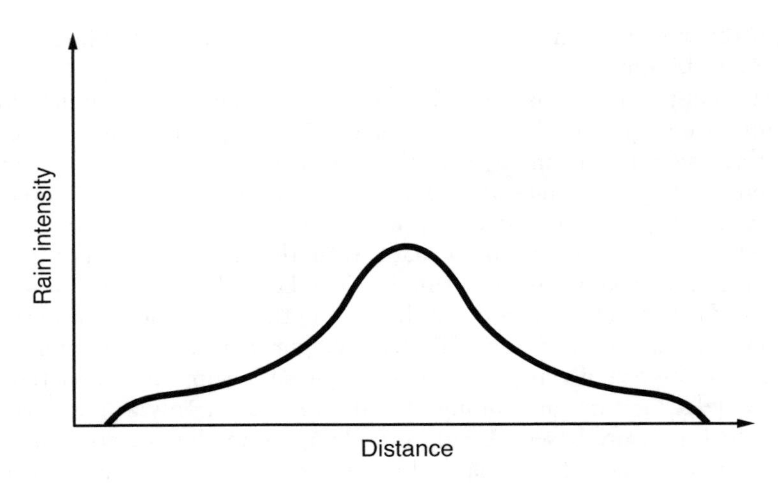

FIGURE 5.1 The *Mexican hat* rain-cell model.

On the fringe of the storm, there is the low-level, fairly widespread rainfall. As you come nearer to the center the intensity increases until you reach the very heaviest rain, which is of a short duration, at the center of the cell. Thus the extent of the high rain rate, and hence the high-attenuation center of the storm, is of very limited dimensions. Hence you cannot directly apply rain-rate data gathered at a particular location (point rain rate) along the complete path length and you must use some form of averaging.

One group of methods characterizes the statistical rain profile simply by a reduction coefficient derived from the spatial autocorrelation function of rainfall [8]. A subset of these derives a reduction coefficient from measurements using rapid-response rain gauges spaced along a line [9,10], or from a semiempirical law [11]. Multiplying point rain rate by this reduction coefficient gives the equivalent path-averaged rain rate.

It follows from the above description of the storm cell that the low rain rate on the edge of a cell occurs for larger percentage times than the high rates, and that when a storm cell crosses a microwave path, these low rain rates are a true representation of what happens along a large part of the path. Thus, the shorter the path and larger the percentage time of interest, the closer the reduction coefficient approaches unity. This is seen in Figure 5.2, [9], which relates the coefficient to path length and percentage time in the United Kingdom. For the 10^{-1}% of time characteristic, the coefficient is unity for path lengths up to 8 km long and then starts to roll off. For 10^{-2}% of time, it is unity only for path lengths up to 1.5 km.

You can see a characteristic for 10^{-4}% of time, but use this with great caution. In the original set of curves were two curves shown for this percentage time—an

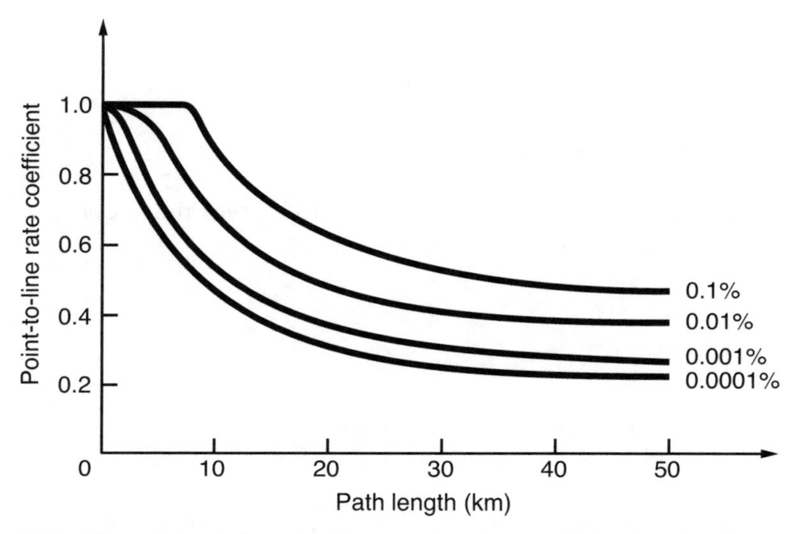

FIGURE 5.2 The variation in the ratio of line-to-point rain rate with hop length and percentage time. (Point rain-rate data derived from a 10-sec response-time gauge.)

upper and a lower limit. We have replaced these with a mean value curve, although the variability of the rain rate for the worst five minutes in the year represented by this curve limits its usefulness.

A commonly used alternative to the above approach applies a reduction coefficient to the actual path length to yield an equivalent path length over which the rain intensity may be assumed to be constant [12].

We now examine two of the methods and then carry out comparative calculations for a United Kingdom location. The first method, described in [9], is commonly used in the United Kingdom where rain-rate data are widely available. The second method [5] is directed at situations in which reliable rain-rate data are not available.

5.5 THE PREDICTION OF RAIN ATTENUATION

5.5.1 The United Kingdom Model

The steps in this model are:

1. Obtain the point rain-rate data for the percentage times of interest with a 10-sec integration time.
2. Determine the point-to-line rain-rate coefficient $C_{P\text{-}L}$ for the percentage times of interest and the true path length from Figure 5.2.

3. Compute the required line rain rates R_L where

$$R_L = R_P \cdot C_{P-L} \quad (5.1)$$

where R_B is the point rain-rate.

4. Compute the specific attenuation, γ_R (dB/km) for the frequency, polarization, and rain rate of interest, where:

$$\gamma_R = kR_L^{\alpha} \quad (5.2)$$

and

R_L is the line rain rate in mm/hr.
k and α are regression coefficients found in Table 5.1.

5. Compute the path attenuation for the given percentage time, given by

$$A = \gamma_R \cdot L \quad (5.3)$$

6. Repeat steps "1" to "5" for any other percentage times of interest.

5.5.2 The CCIR Model

Take the following steps:

1. Obtain the rain rate $R_{0.01}$ exceeded for 0.01% of the time (with an integration time of 1 min). If this information is not available from local sources of long-term measurements, obtain an estimate from the information in § 4 of Report 563.
2. Compute the specific attenuation, γ_R(dB/km) for the frequency, polarization, and rain rate of interest.

where

$$\gamma_R = kR^{\alpha} \quad (5.4)$$

and

R is the rain rate in mm/hr.
k and α are regression coefficients found in Table 5.1.

TABLE 5.1
Regression Coefficients for Estimating Specific Attenuation

Frequency (GHz)	k_H	k_V	α_H	α_V
1	0.0000387	0.0000352	0.912	0.880
2	0.000154	0.000138	0.963	0.923
4	0.000650	0.000591	1.121	1.075
6	0.00175	0.00155	1.308	1.265
7	0.00301	0.00265	1.332	1.312
8	0.00454	0.00395	1.327	1.310
10	0.0101	0.0087	1.276	1.264
12	0.0188	0.0168	1.217	1.200
15	0.0367	0.0335	1.154	1.128
20	0.0751	0.0691	1.099	1.065
25	0.124	0.113	1.061	1.030
30	0.187	1.167	1.021	1.000
35	0.263	0.233	0.979	0.963
40	0.350	0.310	0.939	0.929
45	0.442	0.393	0.903	0.897
50	0.536	0.479	0.873	0.868
60	0.707	0.642	0.826	0.824
70	0.851	0.784	0.793	0.793
80	0.975	0.906	0.769	0.769
90	1.06	0.999	0.753	0.754
100	1.12	1.06	0.743	0.744
120	1.18	1.13	0.731	0.732
150	1.31	1.27	0.710	0.711
200	1.45	1.42	0.689	0.690
300	1.36	1.35	0.688	0.689
400	1.32	1.31	0.683	0.684

Source: CCIR Report 721–3, page 230, Annex to Volume 5, by permission of the ITU.

3. Compute the effective path length L_{eff} of the link by multiplying the actual path length L by a reduction coefficient r. An estimate of this factor is given by

$$r = 1/(1 + L/L_o) \tag{5.5}$$

where

$$L_o = 35\exp(-0.015R_{0.01}) \tag{5.6}$$

4. Give an estimate of the path attenuation exceeded for 0.01% of the time by

$$A_{0.01} = \gamma_R L_{\text{eff}} = \gamma_R L_r \tag{5.7}$$

5. Deduce attenuations exceeded for other percentages of time P in the range 0.001% to 1% from

$$A_p / A_{0.01} = 0.12p^{-(0.546 + 0.043 \log P)} \tag{5.8}$$

This formula gives factors of 0.12, 0.39, 1, and 2.14 for 1%, 0.1%, 0.01%, and 0.001%, respectively.

5.5.3 Comparative Calculation

For the purpose of this calculation, we assume that we need to predict the path attenuation from rain on a 20-km, 15-GHz horizontally polarized link near Manchester in the United Kingdom.

Using the United Kingdom model:

1. Use the rain rates for this location with an integration time of 10 sec that are 12, 32, and 70 mm/hr for 0.1, 0.01, and 0.001% of time.
2. Use the point-to-line rain-rate coefficients obtained from Figure 5.1 that are 0.63, 0.5, and 0.37 for the three given percentages of time.
3. Use the line rain rates (mm/hr) at the three given percentages of time that are therefore

$$12 \times 0.63 = 7.56, \; 32 \times 0.50 = 16, \text{ and } 70 \times 0.37 = 25.9$$

4. The values of k and α at 15 GHz horizonal polarization are 0.0367 and 1.154. Hence use the specific attenuations (dB/km) at the three given percentages of time that are therefore

$$0.0367 \times 7.56^{1.154} = 0.38$$

$$0.0367 \times 16^{1.154} = 0.899$$

$$0.0367 \times 25.9^{1.154} = 1.569$$

5. Multiply these values by the path length to obtain the required path attenuations:

$$7.6 \text{ dB}, 17.98 \text{ dB and } 31.38 \text{ dB}$$

Using the CCIR model:

1. From CCIR Report 563, obtain a value of $R_{0.01}$ of 30mm/hr.
2. Use the specific attenuation for 15-Ghz horizontally polarized:

$$0.0367 \times 30^{1.154} = 1.859 \text{ dB/km}$$

3. Then determine the reduction coefficient of path length:

$$L_o = 35 \exp(-.015 \times R_{0.01}) = 12.42$$

$$r = 1/(1 + 20/12.42) = 0.383$$

4. Determine the path attenuation for 0.01% of time:

$$1.859 \times 20 \times 0.383 = 14.24 \text{ dB}$$

5. Using the multiplying factors determined, the path loss for 0.1 and 0.01 percentage of time compute:

$$14.24 \times .39 = 5.55 \text{ dB and } 14.24 \times 2.14 = 30.47 \text{ dB, respectively.}$$

Comparing the two sets of results and taking into account the variability of the rain rates with the estimates used in the CCIR method, the results obtained show reasonable agreement. The big advantage of the CCIR model is that you can use in locations where detailed rain-rate data are not available.

5.5.4 Polarization Scaling

If rain attenuation figures for a path of given parameters and polarization have been obtained by calculations similar to those above, and you wish to determine the attenuation values using the alternative polarization, you have two possible approaches. First you can repeat the calculations using the values of the regression coefficients k and α given in Table 5.1 for the alternative polarization, or you can use the two scaling expressions found in Report 338 of the CCIR.

These are

$$A_V = 300A_H/(335 + A_H) \text{ dB} \tag{5.9}$$

and

$$A_H = 335A_V/ (300 - A_V) \text{ dB} \tag{5.10}$$

5.6 PATH DIVERSITY

Height diversity used to reduce multipath fading on a link does not improve rain attenuation. However, path diversity in which the same traffic is carried on two paths separated by several kilometers may reduce fading from this cause. Experimental data in the range of 20 to 40 GHz obtained in the United Kingdom [13] shows that you can achieve an improvement in link reliability by using switched path diversity. The following observations were made:

The diversity gain in decibels:

- Tends to decrease as the path length increases beyond 12 km for a given percentage time and path separation;
- Is generally greater for a spacing of 8 km than for 4 km, although a further increase to 12 km provides no further improvement;
- Is not significantly dependent on frequency in the 20- to 40-GHz range for a given path geometry;
- Ranges from around 2.8 dB at 0.1% of the time to 4 dB at 0.001% for a spacing of 8 km and path lengths of 12 km.

Although this is a worthwhile improvement, it seems better overall to reserve that part of the spectrum in which significant rain fading is experienced for short spur routes and the like.

5.7 MOLECULAR ABSORPTION

The absorption from oxygen and water vapor in the atmosphere is an additional factor to consider at certain frequencies, the specific attenuation in decibels/kilometer being shown in the curves of Figure 5.3 for the 3- to 350-GHz frequency range.

The first peak in the water-vapor characteristic at 22 GHz poses no particular problems, since links operating at this frequency are limited in length by the need to avoid outage from rain attenuation, so the additional 0.17 dB/km offers no problems. Calculation shows that with a link as short as 10 km, the rain attenuation for 10^{-3} percentage of time is around 40 dB, and the 1.7 dB from absorption negligible.

The water-vapor absorption lines at 183 and 325 GHz pose considerable problems for any system of length greater than a few hundred meters. It is interesting to see what role may appear for these frequencies in the future.

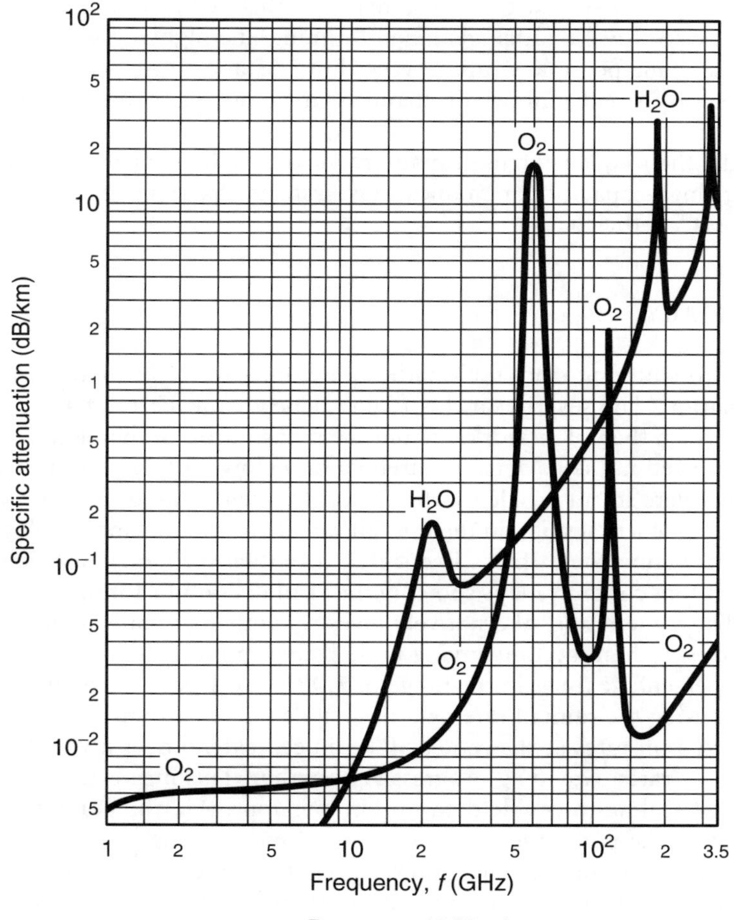

FIGURE 5.3 Specific attenuation from atmospheric gases (reproduced from CCIR Report 719–8 by ITU permission).

You might think that the region around 60 GHz would prove difficult but this is not the case. Three particular uses spring to mind. The first is a military use in which the 15-dB/km specific attenuation offers the chance to use short-range links, knowing that the traffic on these is secure against anyone trying to intercept it at distances of 4 km or over.

The second use is in the area of traffic information or route guidance, in which the narrow beams possible with small antennas at this frequency allow passing information traffic to travel on a single lane of a multilane highway. Further, with frequency reuse made possible within a very short distance from the high-absorption loss, only a very small number of channels are required to provide the required coverage.

Finally, the frequency reuse capability makes the band attractive for building-to-building, high-capacity communications systems that meet the ever-increasing need for spectral efficiency.

5.8 SUMMARY

In addition to suffering from multipath fading, microwave links are susceptible to rain attenuation. The severity of this fading depends on the rainfall rates, the frequency on which the link is working, and the polarization of the signal. In northern Europe, you need not consider rain attenuation below 11 GHz, but in regions suffering from severe storm problems, the critical frequency can be as low as 5 GHz.

You can sometimes obtain the rainfall rate data from the meteorological office of the country concerned. If this is not the case, then recourse must be made to a CCIR report that contains maps showing the rainfall contour for 0.01 percentage of the time. The same report alternatively gives a number of maps divided into rain-climatic zones with an accompanying table listing rainfall intensity exceeded for a number of percentage times in the range of 1.0 to 0.001 percentage of time.

When using the rainfall rate data in a model, take account of the fact that the high-intensity rain rates at the center of a rain cell extend over a very limited area, with the lower rates extending progressively farther. This means that you must use some form of scaling to enable data use. In one approach, the point rain-rate data are scaled with the percentage time of interest, and the length of the radio path is considered. This determines a line rain rate that you can consider applying along the full length of the path when carrying out the attenuation calculation. An alternative approach derives a scaling factor that you apply to the path to obtain a pseudo path length over which the point rain-rate data can be assumed applicable.

In situations where detailed rain-rate data are not available, you can use a CCIR model. This model is based on the rain-rate data for 0.01 percentage of time, which is used in the pseudo path length model to obtain the attenuation for this percentage of time. Scaling factors are applied to this attenuation to obtain attenuation values for other percentage times in the range of 1.0 to 0.001.

Note that multipath fading and rain attenuation are mutually exclusive events. This is due to the strong winds associated with heavy rain cells that break up any atmospheric layering.

Path diversity can be successful in reducing the effect of rain fading. In this form of diversity, the traffic is carried over two routes laterally separated by a few kilometers, with the best signal being selected at the end of the route.

Molecular absorption, in particular that from oxygen, is of interest for specific uses. In the 60-GHz band, the absorption results in a specific attenuation of around 15 dB/km. Although on the face of things this seems like a band to be avoided, there are particular services that do not require the facility of long-distance transmission and can employ the small reuse distance for efficient use of the spectrum.

References

[1] Morita, K., et al., "Radio Propagation Characteristics due to Rain in the 20-GHz Band," *Rev. Elec. Comm. Labs.* Vol. 22, No. 7–8, 1974, pp. 619–632.

[2] Chu, T. S., "Rain-Induced Cross-Polarisation at Centimetre and Millimetre Wavelengths," *BSTJ,* Vol. 53, No. 8, 1974, pp, 1557–1579.

[3] Laws, J. O., and Parsons, D. A., "The Relation of Raindrop Size to Intensity," *Trans Amer. Geophys. Union,* Vol. 24, 1943, pp. 452–460.

[4] Marshall, J. S., and Palmer, W., "The Distribution of Raindrops With Size," *J. Meteorol,* Vol. 5, August 1948, pp. 165–166.

[5] CCIR, *Reports of the CCIR 1990.* ITU, Annex to Vol. 5, Geneva 1990, Report 338–6, pp. 371–372.

[6] CCIR, *Reports of the CCIR 1990,* ITU, Annex to Vol. 5, Geneva 1990, Report 563–4.

[7] Fedi, F. "Prediction of Attenuation due to Rainfall on Terrestrial Links," *Radio Sci.,* Vol. 16, No. 5, 1981, pp. 731–743.

[8] Morita, K., and Higuti, I., "Prediction Methods for Rain Attenuation Distributions of Micro and Millimetre Waves," *Rev. Elec. Comm. Labs* (Japan), Vol. 24, No. 5–6, 1976.

[9] Harden, B. N., et al., "Estimation of Attenuation by Rain on Terrestrial Radio Links in the United Kingdom at Frequencies from 10 to 100 GHz," *IEEE J. Microwaves, Optics and Acoustics,* Vol. 2, No. 4, 1978, pp. 97–104.

[10] Crane, R. K., "Prediction of Attenuation by Rain," *IEEE Trans. Comm.,* Vol. Comm-28. No. 9. 1980, pp. 1717–1733.

[11] Battesti, J., and Boithias, L., *6th Colloquium on Microwave Commnications,* Budapest, Hungary, 1978.

[12] Lin, S.H., "A Method for Calculating Rain Attenuation Distributions on Microwave Paths," *BSTJ,* Vol. 54, 1975, pp. 1051–1086.

[13] Harden, B.N., et al., "Attenuation Ratios and Path Diversity Gains Observed in Rain on a Network of Short Terrestrial Links at Frequencies Near 11, 20, and 36 GHz," IEEE Conference Proc., No. 169, 1978.

CHAPTER 6

SYSTEM OUTAGE MODELING

6.1 INTRODUCTION

By agreement, any multichannel radio link that may form part of an international circuit (and in practice this usually means all links) must operate to standards recommended by the CCIR. These standards are defined in the form of performance objectives for hypothetical reference links 2,500 km long. The makeup of these links is described in *Recommendations* to be found in Volume 9 of the CCIR texts, Recommendation 393 applying to analog systems, and Recommendation 594 being that for digital equipment.

In this chapter we discuss the CCIR performance objectives and how to scale them down for application to practical hop lengths; the concept of worst-month; and the meaning of several terms used in the performance objectives. We then examine available performance-prediction models and make critical assessments of their effectiveness.

6.2 CCIR PERFORMANCE OBJECTIVES FOR ANALOG SYSTEMS

These performance objectives are reproduced from CCIR Recommendations 393–4, *Unanimously Recommends 1,* by ITU permission. The allowable noise power in any telephone channel on a hypothetical reference circuit (HRC) (Figure 6.1) for frequency-division multiplex radio-relay systems is defined in the following way:

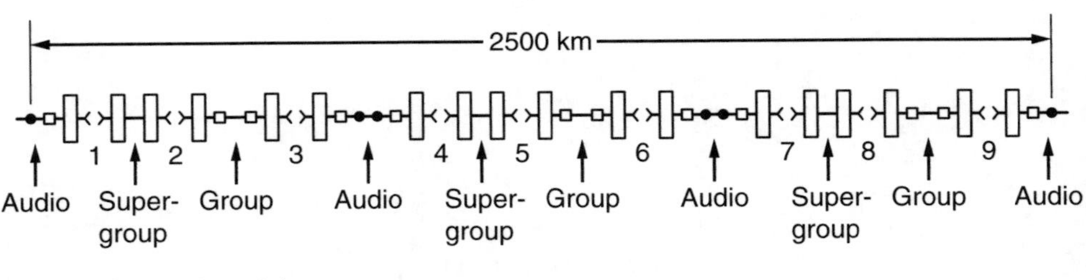

FIGURE 6.1 HRC for radio-relay systems using frequency-division multiplex with a capacity of more than 60 telephone channels per radio-frequency channel (reproduced from CCIR Recommendation 392 by ITU permission).

The noise power at a point-of-zero relative level in any telephone channel should not exceed the following:

- 7,500 picowatts at point-of-zero relative level psophometrically weighted (pW0p) 1-min mean power for more than 20% of any month;
- 47,500 pW0p psophometrically weighted 1-min mean power for more than 0.1% of any month;
- 1,000,000 pW0 unweighted (integration time 5 mS) for more than 0.01% of any month.

Before proceeding any further, we must look at some terms used above.

Psophometrically weighted means that in making the measurement, we must consider the characteristics of the human ear.

Multiplexing concerns the way in which many voice channels are impressed onto a single carrier to meet the traffic-capacity requirements. In analog systems, the system bandwidth is divided into a number of slots, each slot carrying one voice channel. Thus a typical high-capacity analog system can carry 960 voice channels on a single radio frequency channel. Such a multiplexing arrangement is known as *frequency-division multiple access.*

In a digital system, instead of allocating a frequency slot to each voice channel, a time slot is allocated. Thus the high-capacity digital bit stream comprises many channels, each channel confined to its own time slot, the slots accessed sequentially. This arrangement is known as *time-division multiple access* (TDMA).

A *hypothetical reference circuit* (HRC) is a tool used when considering the performance of microwave systems. The HRC is 2500 km long and is divided into a number of sections. The overall performance target for circuit is defined, and this can then be divided into targets for the performance of individual radio links so that when they operate in tandem, they meet the overall target performance, whatever the overall system length.

Worst-month we already defined in Chapter 3, but at this point we need a better understanding of the term. The CCIR *Recommendation* refers to performance in *any month*. Since not all months suffer the same degradations, the performance limits must be met for the most severe month. The worst-month concept can best define this from examination of a period for which copolar fading records are available.

The degradations in system performance result from adverse propagation conditions. For analog systems, the original basis for the worst-month concept, we can directly equate the performance degradations to the reduction in received signal level caused by multipath fading, and, in the case of higher frequencies, by rain attenuation as well. (These two degradations are independent of each another, and their durations must be summed.) This relationship between performance degradation and signal-level reduction results from the AGC action of the receiver. As the signal level into the receiver falls, the AGC system compensates by increasing the receiver gain. As explained earlier, however, this is at the expense of increasing the front-end noise of the system and thus degrades the signal-to-noise ratio. Once this exceeds the previously mentioned thresholds, the system is judged as degraded.

The system designer therefore needs to know what signal attenuation to expect at any location for the relevant percentages of time. To obtain this you refer to the worst-month fading characteristics. This characteristic is usually derived by plotting the cumulative distributions (CDs) of signal attenuation on a log-linear basis for twelve consecutive months, on the same pair of axes. The envelope curve for the complete set is then the worst-month characteristic (Figure 6.2). Note that for simplicity, this figure shows only four months' characteristics.

If many years of data are available, then the worst-month characteristic is the mean envelope curve for the mean monthly characteristic of each of the twelve-month sets. If, on the other hand, no data are available, then for the case of multipath fading, you can approximate the curve by calculating the fading for 0.1% of the worst-month (the $F_{0.1}$ figure) by using the multipath prediction model (found later in this chapter) and then drawing a 10-dB/decade characteristic through this point. This produces a usable worst-month characteristic for all attenuations more than around 15 dB. For higher frequencies, calculate the attenuation produced by heavy

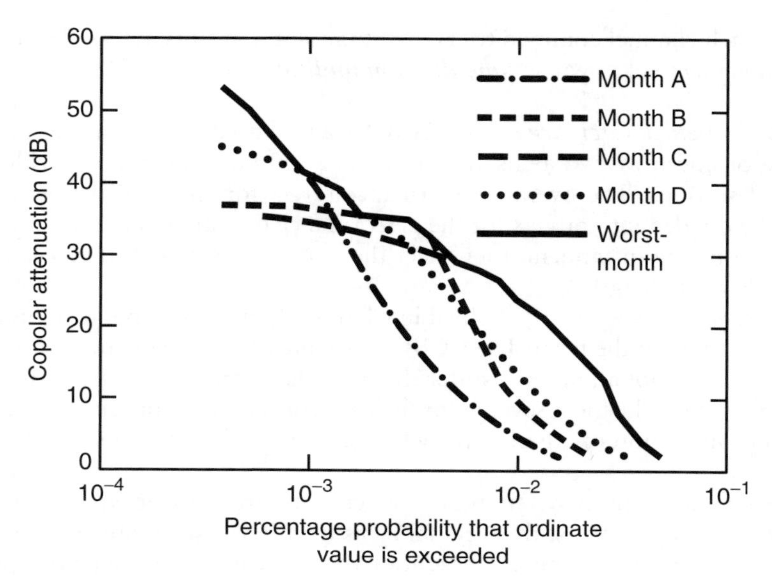

FIGURE 6.2 Worst-month characteristic.

rainfall and use it to derive an additional outage factor, using either local rainfall data or the coarser information from the CCIR reports on this subject.

The worst-month characteristic, because of its construction, may represent fading from different months at different percentage times. This fact can prove puzzling to many people who point out, quite rightly, that it does not represent a real situation. From the planner's point of view, however, it is a useful tool, the utility of which is fully justified. Let us assume that someone is planning a link, for which the worst-month characteristic is that shown in Figure 6.2. Let us further assume that the link is 50 km long. The first step is to calculate the percentages of the month that indicate the performance targets as applied to his link. Since the noise powers of tandem links are cumulative, you can scale the figures of 20, 0.1, and 0.01% down linearly from the HRC targets by using the relationship

$$\text{Link percentage} = \text{HRC percentage} \times L/2{,}500 \tag{6.1}$$

and so that they become 0.4, 2.10^{-3}, and 2.10^{-4}% respectively.

The next step is to read off the attenuations indicated for these scaled percentage times. An example is 35 dB at 2.10^{-3}% of time. You then examine the characteristics of the radio equipment you intend to use to ensure that these attenuations do not exceed the respective permitted noise powers. If this is the case, then all is well and the link-planning process can proceed. If, however, the target performance is not met, then you must take other measures, such as higher transmission power, increased antenna gain, use of space or frequency diversity. You must consider each alternative on its merits, however, since there may be penalties associated with their use. Higher transmission power, for instance, may involve an unacceptable increase in equipment cost, especially at higher frequencies where the development of high-power, solid-state devices has not progressed very far. There is also an upper limit to the effective radiated power (ERP) permitted by the regulatory authorities. Frequency diversity is spectrally inefficient and may also exceed regulatory guidelines. Space diversity is often the only acceptable solution and does not necessarily involve too high costs.

6.3 MULTIPATH PREDICTION MODELS

6.3.1 The United Kingdom Model

Report 338 of Volume 5 of the CCIR has a basic, narrowband multipath prediction model, with coefficients quoted for several regions of the world. In the United Kingdom, an extended model [1] has been developed, based on data derived from a study of 29 links in the United Kingdom, covering 2 to 37 GHz and with path lengths in the range 7.5 to 75 km. The model developed at British Telecom Laboratories is now in general use in the United Kingdom and has been adopted in a slightly different form by CNET in France. The model (below) gives a prediction of the fading exceeded for 0.1% of the worst month in the average year, using the parameters frequency (GHz), path length (km), ground roughness (defined as the rms value of the ground slopes in milliradians, measured between points separated by 1 km along the path, but excluding the first and last kilometric intervals), and a geographical location factor.

$$F_{0.1} = -28 + 35 \cdot \log d + 8.5 \cdot \log F - 14 \cdot \log R + G \tag{6.2}$$

The values of G, the geographical location factor for the United Kingdom, are shown in Figure 6.3. They were obtained by comparing predicted and measured values of the $F_{0.1}$ for all 29 links used in the study and fine-tuning for minimal prediction error. We can identify physical reasons for the variation in G in different parts of the United Kingdom. In the east, for instance, advection ducting (Ch. 1) is a common occurrence but breaks down, in the presence of high ground causing mixing of the atmospheric layering, before reaching the midlands—hence the 6-dB difference between the two locations.

You can easily derive models such as this if you have reliable fading records for a large number of links with a good mix of hop lengths and frequencies. However, if

FIGURE 6.3 The geographic term of the United Kingdom multipath prediction model.

you need models for developing countries without any fading databases, and possibly without detailed contoured maps to calculate the ground roughness factor, then some other approach is necessary.

6.3.2 CCIR Prediction Models

This subsection is reproduced from CCIR report 338-6 by ITU permission. Methods 1 and 2 below give predictions for narrowband systems in any part of the world. Method 2 is the more accurate, based on tests against measured data.

Method 1 requires only path length d (km), frequency f (GHz), and path inclination ε_p (mrad) and is best suited for planning purposes. Method 2 requires in addition the grazing angle (mrad) of the wave specularly reflected by the average surface under the path and is better suited for detailed link design on specific paths for which the path profile is available. You can also use these methods provisionally for over-water paths (with strong surface reflections) with a modified geoclimatic factor K.

Method 1 for Initial Planning Purposes

Note that for the path location in question, estimate the geoclimatic factor K for the average worst-month from measured fading data for the area if these are available, using this method.

1. Obtain the worst calendar month envelope fading distribution for each year of data. Average these for the cumulative fading distribution for the average worst-month and plot this on log/linear graph paper.
2. From the graph, note the fade depth A_1, beyond which the cumulative distribution is approximately linear, and obtain the corresponding exceedance probability P_1. This linear portion constitutes the large fade depth tail, which can vary up to about 3 or 4 dB/decade in slope about the average Rayleigh value of 10 dB/decade, the amount of this variation depending on the number of years of data contained in the average distribution.
3. Calculate the path inclination ε_p from (6.5) and the grazing angle φ from (6.9) to (6.16).
4. Substitute the coordinates (P_1, A_1) of the "first tail point" into (6.17) along with the values of d, f, $|\varepsilon_p|$ and φ and calculate the geoclimatic factor K.
5. If data are available for several paths in a region of similar climate and terrain, or several frequencies on a single path, obtain an average geoclimatic factor by taking an average of the values of log K.

Note that if such data are not available you can estimate K from the contour maps of figures 8 through 11 of CCIR Report 563–3, showing the percentage of time P_L that the average refractivity gradient in the lowest 100m of the atmosphere is less than –100 units/km, and the following empirical relations:

For overland paths not in high, mountainous regions:

$$K = 10^{-6.5} P_L^{1.5} \tag{6.3}$$

For overland paths in high, mountainous regions:

$$K = 10^{-7.1} P_L^{1.5} \tag{6.4}$$

Choose the month that has the highest value of P_L from the four months for which maps are given in figures 8 to 11 of CCIR Report 563–3. An exception to this is that the February map should not be used in the Arctic since, for some reason, it gives false results.

In mountainous areas for which the data used to prepare the maps, in the above-mentioned report, are nonexistent or very sparse, these maps have insufficient detail and K estimated from (6.3) tends to be an upper bound. Such areas include the mountainous regions of western Canada, the European Alps, and Japan. You can use the adjustment contained in (6.4) until more detailed maps become available.

If the antenna heights h_e and h_r (in meters above sea level or some other reference height) are known, calculate the magnitude of the path inclination $|\varepsilon_p|$ in milliradians.

$$|\varepsilon_p| = |h_r - h_e|/d \tag{6.5}$$

where the path length d is in kilometers.

Calculate the percentage of time P that the fade depth A (dB) is exceeded in the average worst-month from

$$P = Kd^{3.6} \cdot f^{0.89} \cdot (1 + |\varepsilon_p|)^{-1.4} \cdot 10^{-A/10}\% \tag{6.6}$$

Method 2 for Detailed Link Design

Obtain the geoclimatic factor K as indicated for method 1, but replace (6.3) and (6.4) by

For overland paths not in high, mountainous areas:

$$K = 10^{-5.4} \cdot P_L^{1.5} \tag{6.7}$$

For overland paths in high, mountainous regions:

$$K = 10^{-6.0} \cdot P_L^{1.5} \tag{6.8}$$

Obtain the magnitude of the path inclination $|\varepsilon_p|$ as in method 1.

From the profile of the terrain along the path, obtain the terrain heights at 1 Km intervals from one terminal and ending 1 to 2 km from the other. Using these heights, carry out a linear regression with the method of least squares for the linear-equation of the average profile.

$$h(x) = a_0 x + a_1 \tag{6.9}$$

where x is the distance along the path. Calculate the regression coefficients from the relations

$$a_0 = \frac{\Sigma_n xh - (\Sigma_n \times \Sigma_n h)/n}{\Sigma_n x^2 - (\Sigma_n x)^2/n} \tag{6.10}$$

$$a_1 = (\Sigma_n h - a_0 \Sigma_n x)/n \tag{6.11}$$

where n is the number of profile height samples. Using (6.9) calculate $h(0)$ and $h(d)$, the heights of the average profile at the ends of the path, and then, using the following (6.12), calculate the heights of the antennas above the average path profile

$$h_1 = h_e - h(0); h_2 = h_r - h(d). \quad (6.12)$$

For paths where the point of specular reflection is fairly obvious (such as on paths over water, partially over water, or partially over flat, level terrain), the heights above the reflecting surface should be used for h_1 and h_2.

If the path is so rough that it is obvious that the main wave interaction with the ground would be one of diffraction from relatively sharp mountain peaks rather than reflection from relatively flat surfaces (even if extended over distances of only a few hundred meters), it may be meaningless to attempt to determine an appropriate value for the average grazing angle. For such a path, we suggest that you apply method 1 and multiply the estimated value of K by a reduction factor $10^{-0.2}$.

Calculate the average grazing angle (corresponding to a 4/3 Earth radius model a_e = 8,500 km):

$$\varphi = [(h_1 + h_2)/d] \cdot [1 - m(1 + b^2)] \quad (6.13)$$

where

$$m = d^2/[4 \cdot a_e (h_1 + h_2)] \quad (6.14)$$

$$c = |h_1 - h_2|/(h_1 + h_2) \quad (6.15)$$

$$b = 2 \cdot \sqrt{[(m + 1)/3m]} \cdot \mathrm{Cos}[\pi/3 + 1/3 \cdot \mathrm{ArcCos}\{3c/2\sqrt{3m}/(m + 1)^3\}] \quad (6.16)$$

In calculating the coefficients m and c the variables d, h_1, and h_2 must be in the same units. The grazing angle φ is in the desired units of milliradians if h_1 and h_2 are in meters and d is in kilometers.

Calculate the percentage time P that the fade depth A(dB) is exceeded from

$$P = Kd^{3.3} \cdot f^{0.93} \cdot (1 + |\varepsilon_P|)^{-1.1} \varphi^{-1.2} \cdot 10^{-A/10} \% \quad (6.17)$$

Note that the values of P obtained in methods 1 and 2 apply only to narrow-band systems. They are considered valid for fade depths greater than about 15 dB or the value exceeded for 0.1% of the worst month, whichever is greater.

Although the above procedures may seem complex, they arise from the need for the CCIR to produce a common prediction method for use worldwide, instead of the many alternatives previously offered in Report 338. Tests against CCIR data-banks show the methods' validity for all areas. For simplicity, we abbreviate the above text slightly and we recommend that you refer to CCIR Report 338 before attempting to use these two methods.

6.4 CCIR PERFORMANCE OBJECTIVES FOR DIGITAL SYSTEMS

We reproduce these performance objectives from CCIR Recommendation 594–2.

"Unanimously Recommends," 3, by ITU permission.

The allowable bit-error ratio at the 64-Kbps output of the 2,500-km hypothetical reference digital path (HRDP), Figure 6.4, should not exceed

- 1×10^{-6} during 0.4% of any month;
- 1×10^{-3} during more than 0.054% of any month.

Note that in addition, the total errored seconds should not exceed 0.32% of any month.

Note several aspects of these objectives:

- The error ratios are referred to as a 64-Kbps circuit, and because of the way in which errors tend to bunch, do not directly correlate this with the error ratio at the 140-Mbps level [2].
- The allowances should not be directly scaled on the basis of hop length, as was the case for analog systems, since once there is an error within a particular timeslot, additional errors in that slot cannot further degrade system performance.

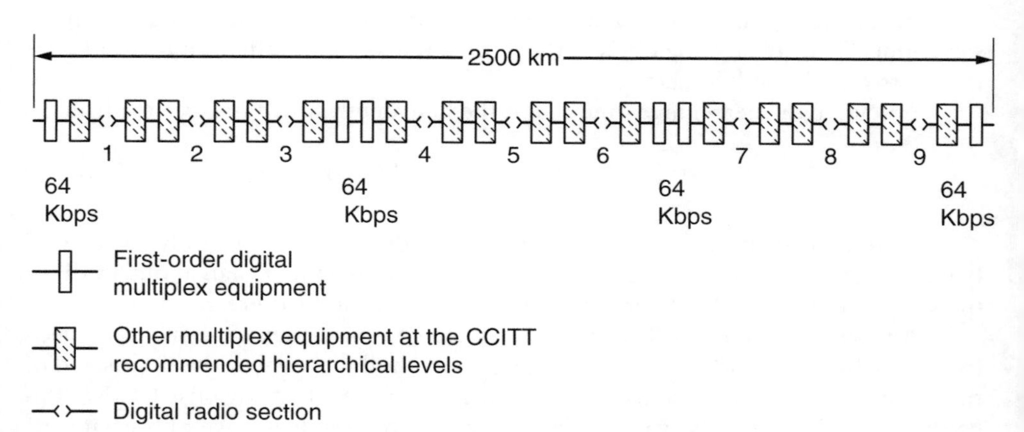

FIGURE 6.4 HRDP for radio-relay systems with a capacity above the second hierarchical level (reproduced from CCIR Recommendation 556–1 by ITU permission).

- A long circuit could have individual hops that were in different climatic zones, and there is a finite probability that they would not both suffer multipath problems at the same time.

The CCIR is considering all these problems.

6.4.1 Availability Objectives

In addition to the performance objectives for analog and digital circuits, we must also meet an availability objective. This objective, which applies to both types of circuit, is defined as follows:

The availability objective appropriate to a 2,500-km hypothetical reference circuit for frequency-division multiplex radio-relay systems and for a 2,500-km hypothetical reference digital path for digital radio-relay systems should be 99.7% of the time.

The concept of unavailability of an analog HRC is that in at least one direction of transmission, one or both of the following conditions occur for at least ten consecutive seconds:

1. The level of the baseband frequencies falls by 10 dB or more from reference level.
2. For any telephone channel the unweighted noise power with an integrating time of 5 ms is greater than 10^6 pWO.

The concept of unavailability for a hypothetical reference digital path is that in at least one direction of transmission, one or both of the following conditions occur for at least ten consecutive seconds:

3. The digital signal is interrupted (i.e., alignment or timing is lost).
4. The bit-error ratio in each second is worse than 1×10^{-3}.

6.5 SYSTEM SIGNATURES

High-capacity digital transmission uses much larger channel bandwidths than those for analog systems. For 140-Mbps systems, this can range from 140 MHz for quadrature phase shift keyed (QPSK) equipment, reducing to around 30 MHz for 64 quadrature amplitude modulated (QAM) equipment. Although much work has been carried out to develop higher order modulation systems, with a corresponding reduction in channel bandwidth, it is the 64-QAM systems that have become the industry standard on the mainline microwave high-capacity systems worldwide. In addition, cofrequency cross-polar (CFCP) working will be used more and more frequently to increase the spectral efficiency of these systems.

Experience on digital systems indicates that the absolute depth of multipath fading is not the prime cause of system outage—multipath distortion is the dominant problem with XPI degradation as a further consideration in the case of CFCP systems. Thus the methods for determining whether an analog system achieves its performance targets are not the same for digital systems.

System signatures can indicate the ability of a particular digital equipment configuration to withstand the distortion effects of a multipath notch close to or within the channel bandwidth, *under defined conditions*. These are derived by setting up a transmitter and a receiver on a test bench and by linking them by way of a multipath simulator. The steps in a measurement sequence are:

1. Adjust the path loss between transmitter and receiver so that the input to the receiver, with no flat fading present, is at the recommended level.
2. Set up the multipath delay to a predetermined value (e.g., 2.3 ns).
3. Set the simulated multipath notch to a frequency (e.g., 50 MHz) below the carrier frequency of the transmitter and increase the notch depth until the measured-error ratio in the system just exceeds a given threshold. Record this value and move the notch in steps until it is 50 MHz above the carrier, making a notch-depth measurement at each step.
4. Plot the locus of the recorded notch depths so that it is the system signature for 0-dB flat fading and 2.3-ns delay.
5. Repeat this procedure for a range of flat-fade levels.
6. Set the multipath delay to a new value and repeat the set of measurements as many times as required. (Derive minimum and nonminimum phase fade signatures if needed.)

In practice, a computer controls the whole procedure for whatever error ratios, notch-frequency intervals, flat-fading levels, and multipath delays are required. Figures 6.5 to 6.7 comprise a typical signature set for 64-QAM equipment, at an error ratio of 10^{-3}. The area under a system signature is a measure of the robustness of the system, so let us examine the factors that affect its width and height.

- The width of the signature is not sensitive to multipath delay when the full system fade margin (defined as the flat-fade that can be tolerated before the system error ratio exceeds 10^{-3}) is available, although this is not the case once flat fading is introduced and the available fade margin is reduced. This is because the notches for a short delay are widely separated in frequency, and so the skirts of notches situated to one side of the channel throw in

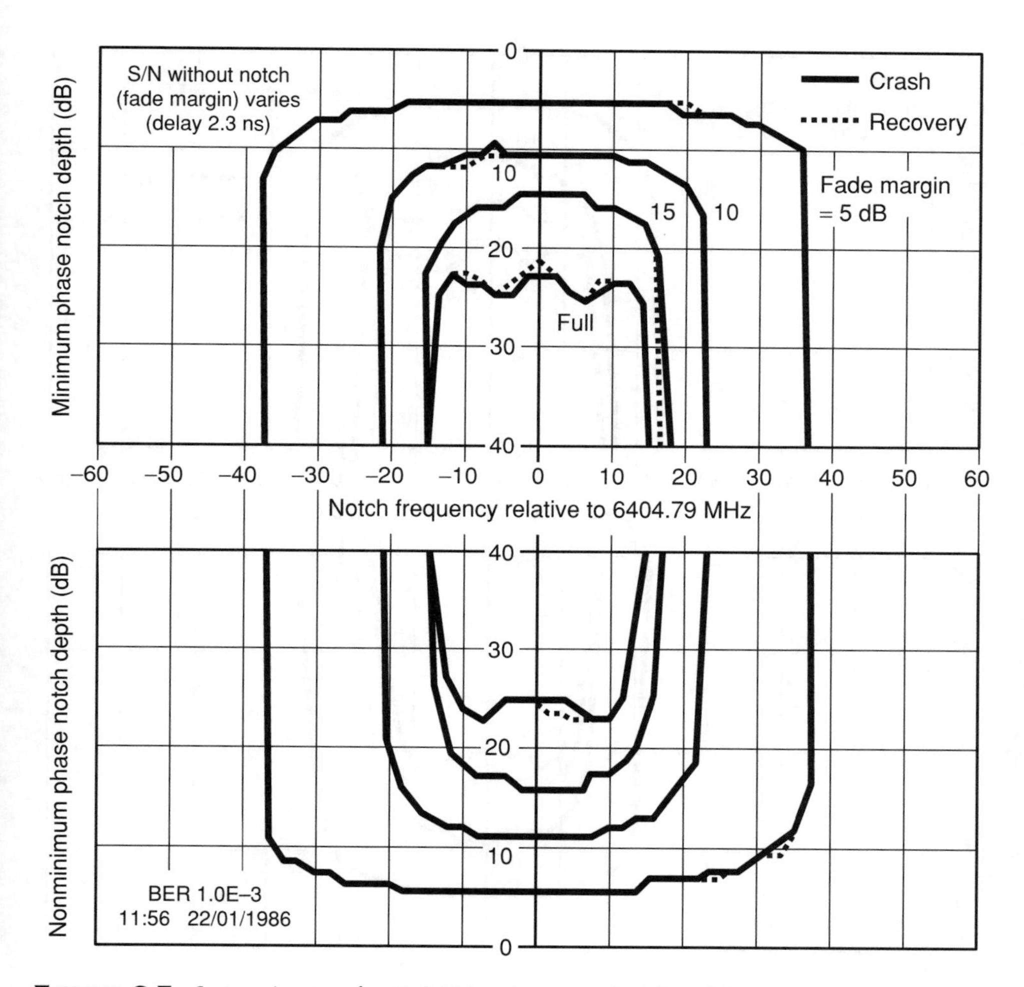

FIGURE 6.5 System signature for 64-QAM equipment and a 2.3-ns delay

what looks very similar to additional flat fading and can cause the system to crash.

- The height of the signature has some delay dependence under full-fade margin conditions.

A very basic use for system signatures is to compare the performance of different sets of equipment. Since the area within the signature is an indication of degraded performance under defined conditions, a simple comparison of these areas

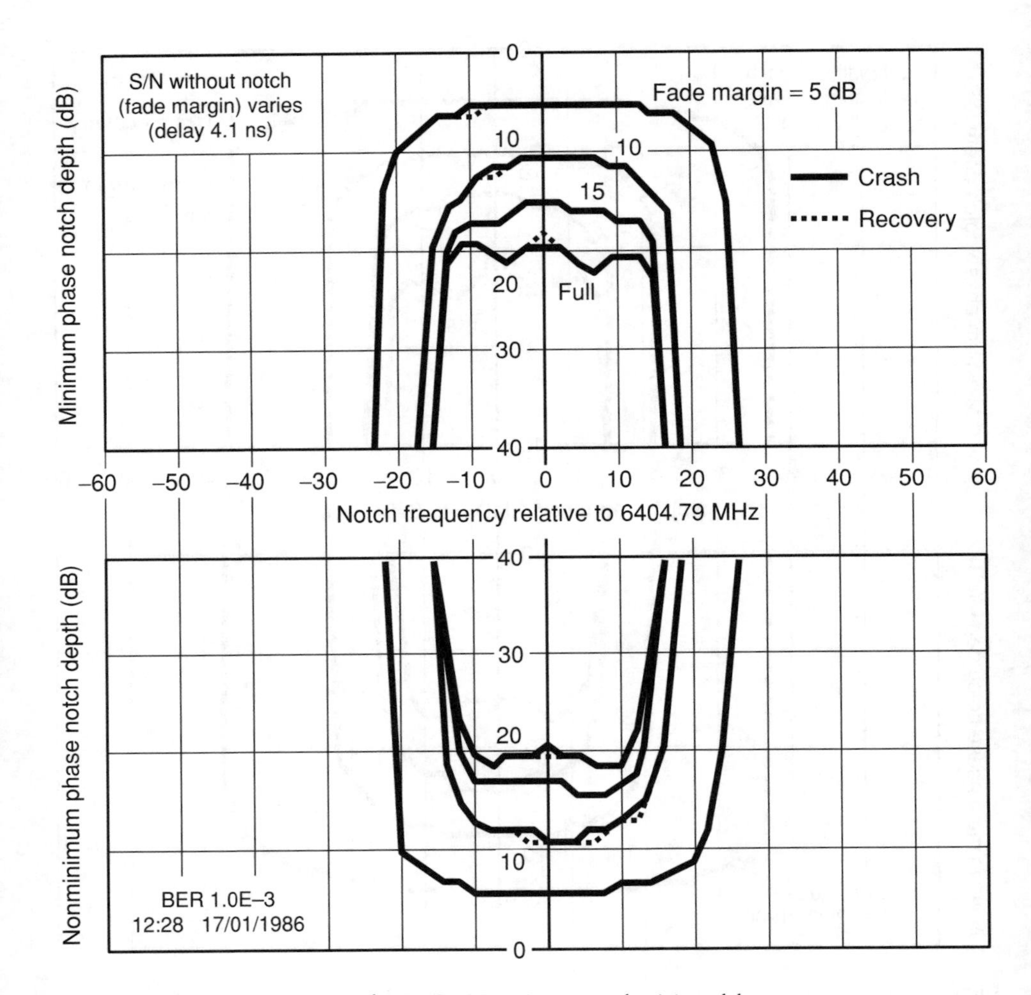

FIGURE 6.6 System signature for 64-QAM equipment and a 4.1-ns delay.

for the two different sets of equipment reveals which of them has the better performance. However, system signatures have a much bigger role to play, since if one has access to a detailed database of fading for an area in which you wish to install a digital link, then as long as the database has information on the occurrence probability of worst-month flat fading, multipath fading, and multipath delays, then you can use the system signatures as a basis for a performance prediction model.

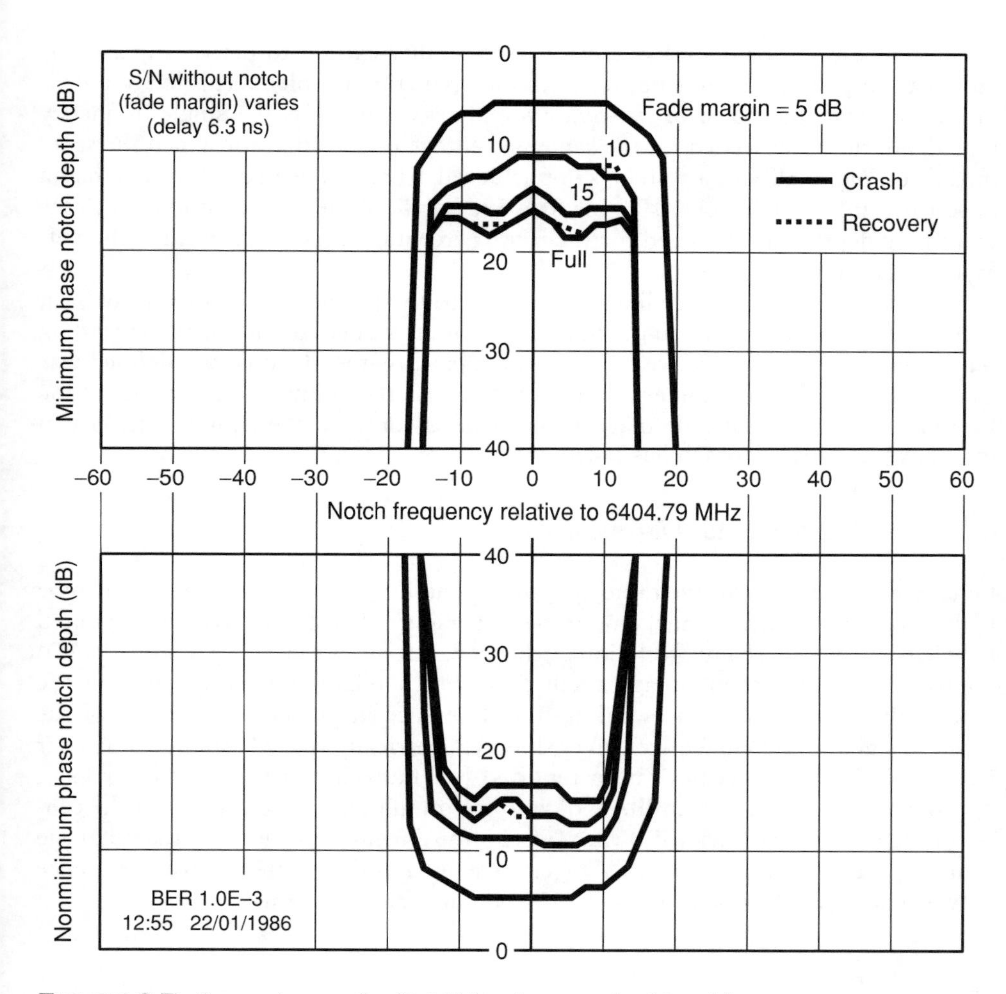

FIGURE 6.7 System signature for 64-QAM equipment and a 6.3-ns delay.

6.6 PERFORMANCE PREDICTION MODELS FOR DIGITAL SYSTEMS

6.6.1 Net-Fade Margin

This model appeared during the early days of digital systems and arose from the recognition that although the analog approach of estimating the probability of multipath fading exceeding the system flat-fade margin to determine system outage could not be used for digital systems, the possibility existed to derive a quasi-flat-fade margin as a very simple approach to the prediction problem. This led to a definition in which the quasi or *net-fade margin* was defined as the single-frequency fade depth that is exceeded for the same percentage time as that for which the specified threshold quality of a system is not achieved. That means that if a system had a bit-error ratio exceeding 1×10^{-3} for, say, 5×10^{-3}% of the time, you looked to see what fade depth was exceeded for the same percentage time. This became the net-fade margin of the equipment.

Although this was a creditable approach to the problem, it had the drawback that in different locations you might arrive at different net fade margins, depending on the mix of flat and selective fading components; that is, the quantity defined was a combination of the equipment characteristics with the fading characteristics of the location, unlike the flat-fade margin that depends only on the equipment. As you might expect, this approach has fallen out of favor.

6.6.2 Inband Amplitude Dispersion

Outage in digital microwave radio systems during frequency-selective fading can be highly correlated with channel amplitude dispersion. This characteristic is exploited to yield a measurement technique, the inband amplitude dispersion (IBAD) approach [3], to system outage prediction, which relates bit-error ratios to the inband dispersion obtained by taking the difference in power (in decibels) of the received signal in two narrow receiver slots symmetrically spaced around the carrier frequency. Good results have been reported [4], but one must query the two-slot approach that would give an IBAD of zero for a multipath notch sitting in the center of the channel bandwidth! If there were a third measurement slot situated at the center of the channel, and the IBAD were redefined as the greatest power difference between any two slots, it would seem to overcome any problem.

6.6.3 Signature-Based Models

The basic concept of such models is that they are based on the interaction between the propagation environment and the signature of the equipment to be used. The model discussed in this subsection is that developed in BT Laboratories and described in the first *European Conference on Radio Relay Systems* in 1986 [5].

Figure 6.8 shows a very simplistic representation of the model in which the equipment has identical signatures for both minimum and nonminimum phase situations and can be represented in the idealized form shown, assuming that it is unaffected by either flat fading or multipath delay. The probability of outage is then:

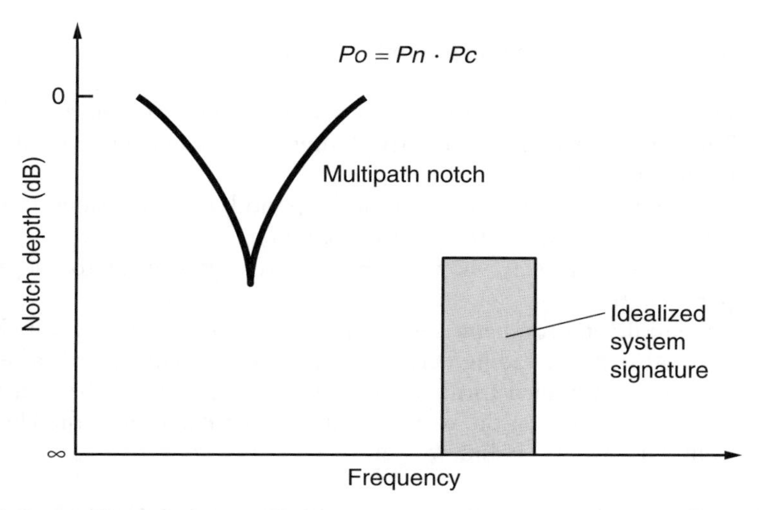

FIGURE 6.8 Simplistic representation of a signature-based performance-prediction model.

$$Po = Pn \cdot Pc \tag{6.18}$$

where

Po is the probability of system outage in the worst-month;
Pn is the probability of occurrence of a multipath notch deeper than the system signature;
Pc is the probability of such a notch falling within the frequency interval occupied by the system signature.

So what is required to evolve this simplistic model into a practical form? Consider the following parameters:

1. Multipath delay, since this affects the signature height;
2. Noise and interference, which can erode the system fade margin and hence affects the signature height and width;
3. Unidentical system signatures for minimum-phase and nonminimum-phase fades, since then the model needs to have information built-in concerning the relative probabilities of these two situations. (However, modern systems do not fall into this category, and you can ignore this complexity.)

Since most administrations have long-term records of single-frequency fading derived from monitoring the AGC lines of analog links, it is important that you use these records in the model if possible.

The basic model is based on a number of assumptions:

1. The flat-fading term has a log-normal probability function.
2. The flat- and frequency-selective fading components are considered statistically independent.
3. An activity factor (the fraction of the period under consideration for multipath activity) scales the joint probability.
4. The multipath delays have probability distribution factor dependent on path length.
5. The fading environment is either in the form of measured distributions of flat and selective fading derived from measurements during a selected period, or in assumed forms of distributions coupled to the fade depth exceeded for 0.1% of the worst-month as calculated with the United Kingdom narrowband fading model.

Based on these assumptions, the model undergoes a triple-integration process in which it examines every combination of flat fade, selective fade, and multipath delay, within a range determined by the input parameters, and calculates the outage probability for each combination.

For example, consider the situation in which the signature is symmetrical about the minimum-phase/nonminimum-phase axis and has an 18-dB height and a 90-MHz width for the combination of a 10-dB flat fade and a 1-ns multipath delay. The model determines the signature parameters, using the input data held in memory. For this situation, there is an infinite set of multipath notches, having different center frequencies and depths, that yield a given selective fade depth (15 dB in this example) in a narrowband slot at the center of the channel. Figure 6.9 shows the situation with three of these notches depicted. Notch A is at the channel center and therefore has a depth equal to the selective fade depth of interest. Notch B is of greater depth, is offset in frequency, and falls within the system signature. Notch C is deeper still but is offset outside the limits of the system signature. Thus the outage probability is determined by summing the occurrence probabilities of all notches that fall within the frequency limits of the channel (F1–F2) and that also have amplitudes exceeding the 18-dB signature height. The total probability is then modified to reflect the occurrence probability of the particular combination of flat-fade and multipath delay.

The model then repeats this procedure until it covers all possible combinations of the three variables (flat fading, selective fading, and multipath delay) within the predetermined limits. Finally, the scaling factor is applied to the calculated outage figure to account for the fact that multipath activity only takes place for a small percentage of time. We abbreviate the explanation of this model's operation for the sake of simplicity and refer you to [5] for the full version. Other models using system

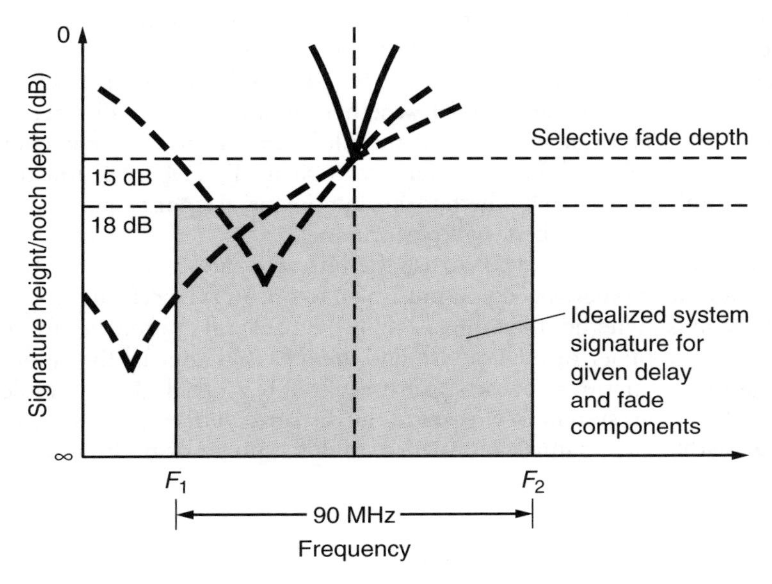

FIGURE 6.9 A set of multipath notches, each producing a 15-dB fade depth at the center of an RF channel.

signatures exist, and although no references are known, you can assume that they use the same basic principles.

6.7 SUMMARY

The traffic carried over high-capacity microwave links is often international, so systems must meet CCIR performance objectives to ensure an acceptable transmission quality at all times. We apply these objectives to 2,500-km-long HRCs, and by scaling down according to path length, we can derive objectives for a link of any length.

To achieve the quality of any circuit under all conditions, individual hops are planned to meet the target performance under worst-month situations, requiring methods for the prediction of multipath activity in any location, as well as a way to link this information to a system performance model.

Narrowband fading prediction models are available for use in many areas, based on long-term fading records. In developing countries, however, these long-term records may not be available, so the CCIR has published methods that you can use in any location worldwide. The first method only requires knowledge of path-length, carrier frequency, and path inclination and is suited for planning purposes.

The second method requires the grazing angle of the wave reflected by the surface under the path and is better suited for detailed link design.

Having obtained a prediction of the worst-month multipath fading distribution by any method, you can establish whether an analog link will meet its performance targets by determining the worst-month fading at the three percentage times of interest, scaled down from the reference path in direct proportion to the hop length. You can also establish whether the equipment's signal-to-noise ratio for the relevant fade depths meets the target performance.

For digital links, however, because the bit-error ratio is not directly linked to the degree of fading experienced you must turn to one of the performance prediction models to establish whether the targets can be met. Several approaches to the prediction of digital system performance exist, and the type that appears the favorite is that based on system signatures. These signatures are a test-bench derived indication of the ability of a digital system to withstand the distortion effects of a multipath notch close to or within the channel bandwidth, under a range of multipath delays, flat fading, and selective fading.

References

[1] Doble, J. E., "Predictions of Multipath Delays and Frequency-Selective Fading on Digital Links in the United Kingdom," *IEE-Colloquium on Modern Techniques for Combating Multipath Interference in Radio, Radar, and Sonar Systems,* London, November 1979.

[2] Casiraghi, U., and Mengali, U., "Relationship Between BER Performance Parameters at 64 Kbit/s and at Radio System Bit Rate," *European Conference on Radio Relay,* Munich, 1986, pp. 360–367.

[3] Martin, A. L., "A New System Measurement Technique," *International Conference on Communications,* Toronto, June 1986.

[4] Martin, A. L., "Dispersion Signatures—Some Results of Laboratory and Field Measurements," *European Conference on Radio Relay,* Munich, 1986, pp. 384–391.

[5] Doble, J. E., "A Flexible Digital Radio System Outage Model Based on Measured Propagation Data." *European Conference on Radio Relay,* Munich, 1986, pp. 64–74.

CHAPTER 7

SYSTEM PLANNING

7.1 INTRODUCTION

We learned about the mechanics of fading on microwave links, how we can use diversity to reduce the severity of the multipath component, and the methods by which we can ascertain that the system will meet performance objectives. In this chapter, we learn how to ensure that the input signal levels to the receivers of a microwave link match the optimal value determined by the manufacturer and how to avoid diffraction fading when the transmission medium becomes very subrefractive. We also learn how to avoid the complexity of working with a curved ray over a curved Earth, by using transformations to a curved ray over a flat Earth or a flat ray over a curved Earth while maintaining the spatial relationships between the ray and Earth's surface.

We also discuss some practical methods used on multihop links to avoid the possibility of interference at repeaters, followed by a planning exercise for an analog link that puts together many concepts that we have already presented in this book. Finally, we discuss the problems that you can encounter when a link passes over water and the countermeasures that you can employ in this area to avoid performance degradation.

7.2 BASIC TRANSMISSION INFORMATION

7.2.1 Free-Space Transmission Loss

If we transmit a signal from an isotropic (unity gain) antenna, then the transmitted energy spreads out uniformly in all directions. So, if we receive this signal some distance away on a second isotropic antenna, then the received signal power is obviously smaller than that transmitted, and the ratio of transmitted power to received power is the transmission loss, usually expressed in decibels. This is called *free-space (FS) transmission loss,* indicating the fact that it is loss that occurs within a perfect and unlimited environment with no outside influence present.

We derive the formula in terms of distance and frequency as follows:

If the power radiated from an isotropic antenna is P_t watts, then the power flux in W/m^2 at a distance of d meters is

$$P = P_t/4\pi d^2 \tag{7.1}$$

For the receiving antenna, the effective aperture area A_e and the gain G are related by

$$A_e = G\lambda^2/4\pi \tag{7.2}$$

thus the received power = $A_e \cdot P$

$$= G\lambda^2 \cdot P_t/4\pi d^2 \cdot 4\pi$$

$$= G\lambda^2 \cdot P_t/(4\pi d)^2 \tag{7.3}$$

Hence the ratio of transmitted power to received power when using an isotropic receive antenna is

$$(4\pi d/\lambda)^2 \tag{7.4}$$

Now d and λ must be in the same units, so if we work in kilometers and use frequency (MHz) rather than wavelength, the ratio becomes

$$(4\pi \cdot d \cdot f/0.3)^2 \tag{7.5}$$

Changing to decibels we arrive at:

$$\text{FS transmission loss} = 32.5 + 20 \cdot \log d + 20 \cdot \log f \tag{7.6}$$

7.2.2 Link Budget

If we are going to plan an analog link, we stated earlier that it will be sensitive to the received signal level, with a requirement to meet an optimal level (determined by the receiver characteristics) under clear-sky conditions to obtain an acceptable performance. Thus the planner, knowing the transmitter output power, needs to know the total loss between the output stage of the transmitter and the front end of the receiver, under clear-sky conditions. This overall calculation is known as the *link budget.*

Assume that the system being considered is a multichannel high-capacity system. There are therefore losses associated with the multiplexing (combining several transmitters onto a single feeder) and demultiplexing (routing the received signals to their correct channels) the feeders at both ends, and the FS transmission loss. In addition there is the antenna gains (those realizable in practice; not the theoretical value obtained from the basic dimensions).

Thus the total loss is

$$L = L_M + L_F + L_{FS} - G_A \tag{7.7}$$

where

L_M = Sum of multiplexer and demultiplexer losses

L_F = Sum of the transmit and receive feeder losses

L_{FS} = FS loss

G_A = Sum of realizable antenna gains

7.2.3 Diffraction Fading and Fresnel-Zone Clearance

The variations in atmospheric refractive conditions can cause changes to the effective Earth's radius or k factor from the median value. If the atmosphere becomes sufficiently subrefractive (i.e., large positive values of RRI gradient), the ray paths are bent downward such that Earth appears to obstruct the direct path between the transmitter and the receiver, causing *diffraction fading*, and the path loss can exceed the FS value by a large amount. This effect is one factor that determines the height at which the antennas are installed.

The diffraction loss severity depends on the type of terrain and the vegetation. For a given ray clearance over the terrain, the minimum diffraction loss is experienced when there is a single knife-edge obstruction, and the maximum for a smooth spherical Earth.

You can approximate diffraction loss over average terrain [1] by

$$Ad = -20h/F_1 - 10 \text{ (dB)} \tag{7.8}$$

Where h is the height in meters of the most significant path blockage above the path trajectory, and F_1 is the radius of the first Fresnel zone, given by

$$F_1 = \sqrt{(d_1 \cdot d_2 \cdot \lambda / d)} \tag{7.9}$$

where

d is the path length.

d_1 and d_2 are the distances from the terminals to the path obstruction.

λ is the signal wavelength.

All parameters are in meters.

The first Fresnel zone is defined as the volume of revolution traced by the ellipse, which is the locus of a point of reflection that provides a half-wavelength delay between the delayed and direct paths between the transmitter and receiver (Fig. 7.1).

The diffraction loss relative to free space (decibel) as a function of normalized clearance h/F_1 is shown in Figure 7.2.

To achieve FS path loss under conditions of extreme subrefractivity, links in the United Kingdom are planned to have 0.6 of the first Fresnel zone free of obstruction under $k = 2/3$ conditions. Other administrations take different approaches to planning; Germany, for instance, plans for complete clearance of the first Fresnel zone under $k = 4/3$ conditions.

7.2.4 Ray Curvature as a Function of RRI Lapse Rate

We can show that the transmission through a spherically stratified medium is such that the transmitted rays are refracted toward the region of higher refractive index (i.e., downward for our situation) with a curvature radius r such that

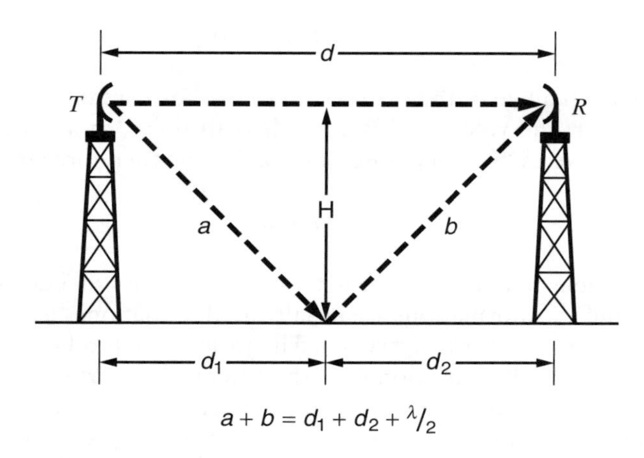

FIGURE 7.1 Definition of the first Fresnel zone.

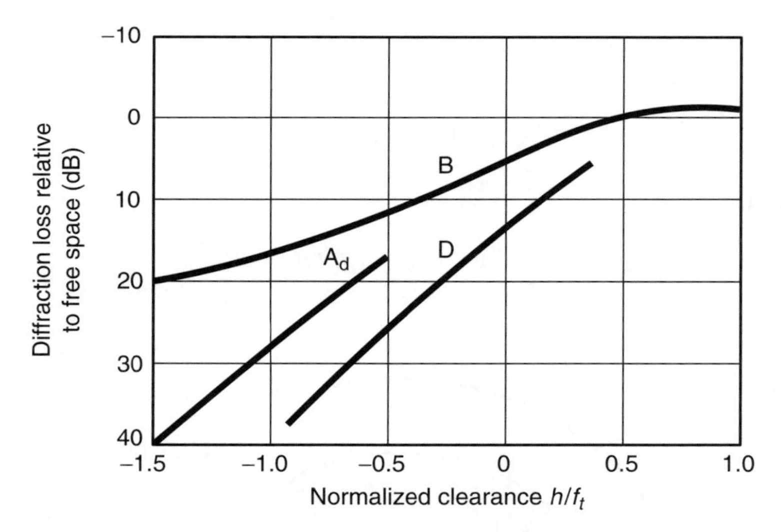

FIGURE 7.2 Diffraction loss for obstructed LOS microwave radio paths (Reproduced from CCIR Report 338–6 by ITU permission).

$$1/r = -1/n \cdot dn/dh \cdot \cos\beta \tag{7.10}$$

where β is the launch angle of the ray relative to the horizontal.

As said earlier, the value of n is very close to unity; in addition, the launch angle β for microwave LOS is very close to zero. We can therefore write the above as

$$1/r = -dn/dh \tag{7.11}$$

Now it is somewhat inconvenient to work in terms of curved rays over a curved surface, and transformations are usually used so that one can use straight-line rays over a curved surface (or the reverse). The former used to be popular for planning, with the gound height variations and the antenna support structures drawn on specially prepared paper, and the signal path represented by a straight line between the antennas. However, the flat Earth approach is now the predominant method.

If r is the true radius of ray curvature a is the curvature radius of Earth, and r_e is the effective ray radius over a flat Earth, then from Figure 7.3 we can see that

$$1/r_e = 1/a - 1/r \tag{7.12}$$

but

$$1/r = -dn/dh \tag{7.13}$$

and therefore

$$1/r_e = 1/a + dn/dh \tag{7.14}$$

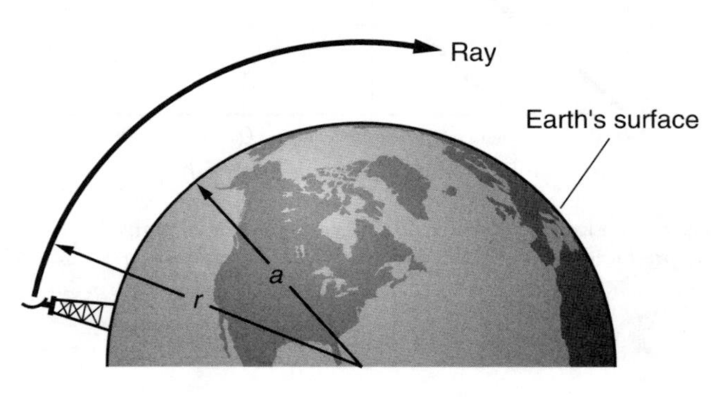

FIGURE 7.3 Curved ray over a curved Earth.

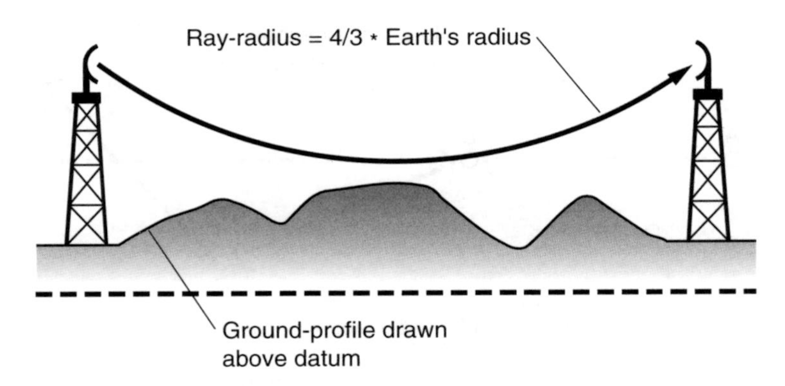

FIGURE 7.4 Curved ray over a flat Earth under median lapse-rate conditions.

By inserting a value of 6,370 km for Earth's radius, we arrive at

$$1/r_e = (157 + dn/dh) \cdot 10^{-6} \quad (7.15)$$

Since dn/dh is usually negative, it follows from (7.15) that the effective ray radius r_e is normally greater than Earth's radius a. The value of r_e/a is known as the k factor, and under clear-sky conditions it has a value of 4/3, a figure used for general calculations (Fig. 7.4).

Note that if dN/dh has a value of –157 N units/km, then r_e becomes infinite. Thus the radius of curvature of the ray is the same as that of the Earth, and we have the long-distance transmission condition known as *ducting*.

7.3 GENERAL PLANNING CONCEPTS

7.3.1 Multihop Links

When planning a system that requires several hops, observe the following rules if possible:

1. The line of shoot of each hop is angularly offset from that of the previous one to avoid overshoot problems. This is known as *dog-legging*. As an extension to this, ensure that no remote antenna (on another link) finds itself on an overshoot path.
2. To help avoid interference situations at repeaters, the polarization of the transmission in a given direction is reversed at each repeater.
3. The traffic is carried on different half-bands on successive hops.

Figure 7.5 depicts these rules put into action. For simplicity's sake this example appears in a very basic form and shows the arrangement for one particular channel.

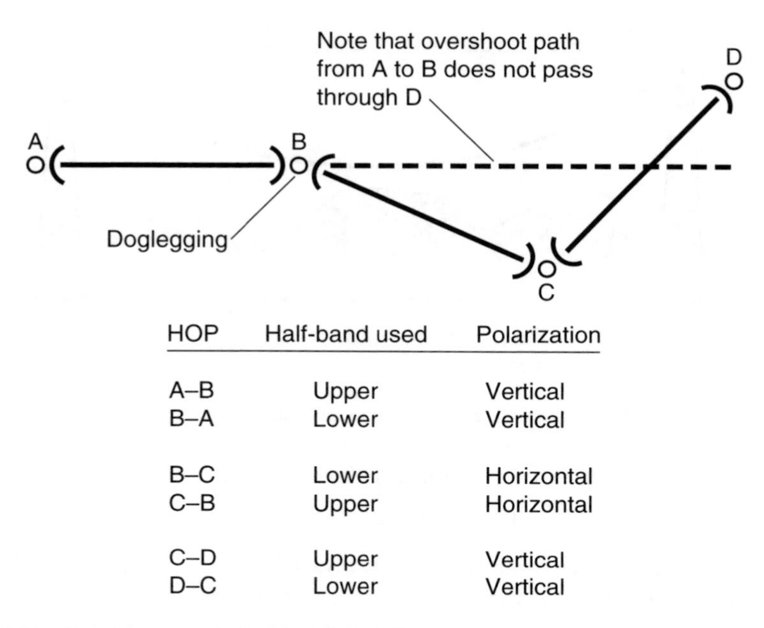

HOP	Half-band used	Polarization
A–B	Upper	Vertical
B–A	Lower	Vertical
B–C	Lower	Horizontal
C–B	Upper	Horizontal
C–D	Upper	Vertical
D–C	Lower	Vertical

FIGURE 7.5 Planning concepts for a multihop link.

To explain the reasons for changing polarization and the half-band used at each repeater, let us look at the situation at repeater B.

Consider first the links between A and B. In both directions, the channel is vertically polarized, but in the A-to-B direction it uses the upper half-band, while in the B-to-A direction it uses the lower half-band. Thus there is no interference.

Next look at the possible problem with energy from the C-to-B hop entering into the B antenna receiving from A. The C-to-B signal is in the upper half-band and is horizontally polarized, whereas the A-to-B signal is in the same half-band but in the opposite polarization. So once again there is no interference.

Finally let us see if there is any problem with any transmitter power from the B-to-A hop leaking into the B antenna of the C-to-B hop. Here the signal transmitted from B (toward A) is vertically polarized in the lower half-band, while the signal received at B (from C) is horizontally polarized and in the upper half-band. Thus the arrangement shown offers full protection against interference at every repeater.

The majority of hops encountered are equipped with space diversity, and it is therefore more usual to use both antennas for transmission and split the channeling arrangements into the format shown in Figure 7.6. This has the benefit of simplifying the multiplexers because of the greater frequency spacing involved, and it also allows work on an antenna without losing the whole system traffic capacity.

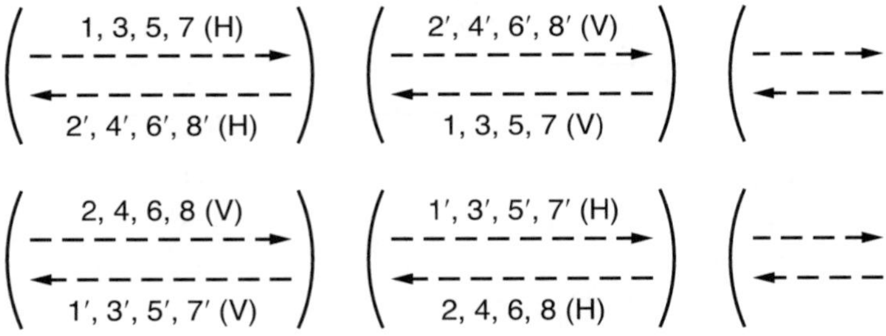

1, 3, 5, 7 (H)
2′, 4′, 6′, 8′ (H)
2′, 4′, 6′, 8′ (V)
1, 3, 5, 7 (V)
2, 4, 6, 8 (V)
1′, 3′, 5′, 7′ (V)
1′, 3′, 5′, 7′ (H)
2, 4, 6, 8 (H)

FIGURE 7.6 Channeling arrangement for a fully equipped, eight-channel half-band system using space diversity.

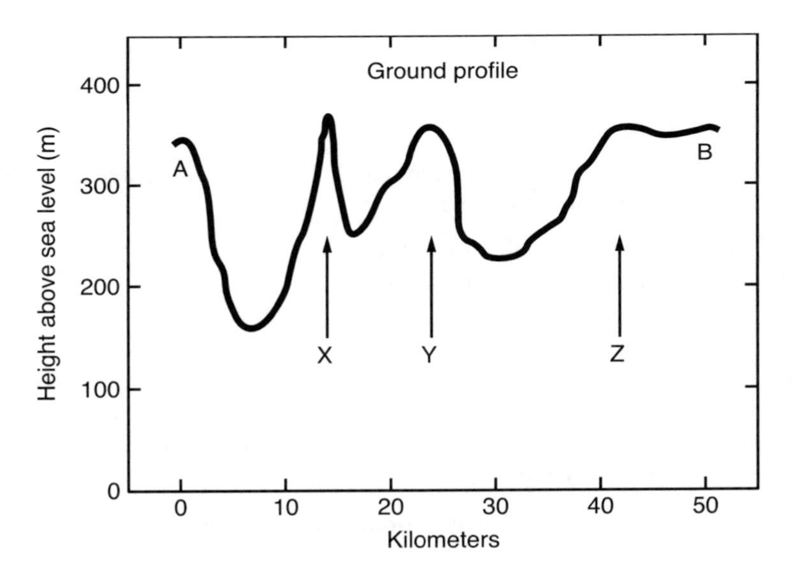

FIGURE 7.7 Ground profile for the planning exercise.

7.3.2 A Planning Exercise

The following planning exercise uses much that you have learned so far. Because of the complexity of predicting outage on digital systems, the exercise concerns an analog link, although many of the steps are common to both systems

Establish a high-capacity analog microwave link between points A and B, separated by 50 km over the terrain shown in Figure 7.7, the path profile. The parameters that apply are shown in Table 7.1.

TABLE 7.1
Link Parameters

Parameter	*Value*
Transmit power	1 W
Optimal receiver input power	–38 dBm
Multiplexer loss	1 dB
Demultiplier loss	1.5 dB
Realizable antenna gain	39 dB
Standard feeder loss	1.8 dB/100m
Low-loss feeder loss	0.7 dB/100m
Receiver flat-fade margin	44 dB
Permitted system outage in the worst-month	3.4×10^{-4}%
Geographical factor	4 dB
Carrier frequency	6 GHz
Ground roughness	10 mrad

Assuming that the antenna at terminal A has to be mounted at 50m above ground level, determine the following:

- The height of antenna at terminal B;
- Whether diversity needs to be used;
- Whether low-loss feeders are required.

Since we need to know the feeder loss before we can draw up the link budget, we will start by determining the height of the antenna at terminal B.

The path profile is drawn on a flat-Earth basis and therefore we need to correct the salient features for the following.

- Earth curvature for the most subrefractive situation we are likely to meet ($k = 2/3$);
- 0.6 of the first Fresnel zone.

The Earth curvature correction C from Figure 7.8 is derived as follows:

If a is the effective radius of the earth and d is the distance of a particular feature from the nearest end of the link, then the earth curvature correction C is derived as follows

$$d^2 + (a - C)^2 = a^2$$

hence

$$d^2 + C^2 = 2aC$$

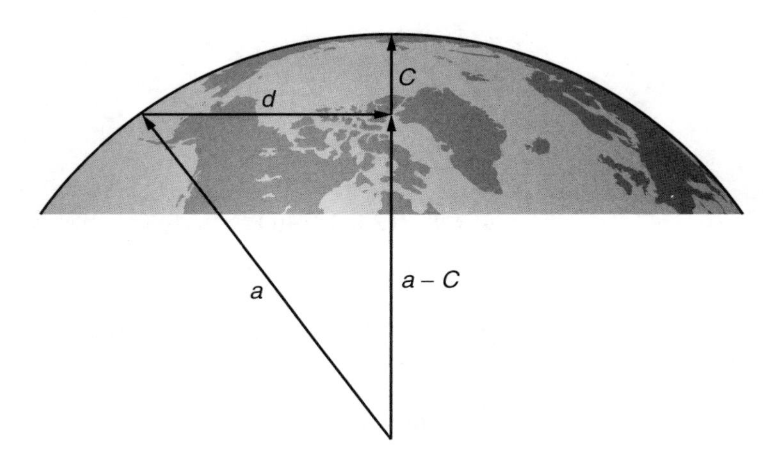

FIGURE 7.8 Earth curvature correction.

but

$$C << d$$

therefore

$$d^2 = 2aC$$

and hence

$$C = d^2/2a \tag{7.16}$$

where a is the effective radius of Earth and d is the distance of the feature from *the nearest* end of the link.

Now the true radius of the Earth is 6,370 km, therefore the value of a under $k = 2/3$ conditions is 4,246 km.

The features on the path profile that you need to consider are X, Y, and Z, where X is 14 km from A, Y is 24 km from A, and Z is 8 km from B.

So:

$$\begin{aligned} C_X &= 14^2/(2 \times 4,246) \\ &= 23\text{m} \end{aligned} \tag{7.17}$$

$$C_Y = 24^2/(2 \times 4{,}246)$$
$$= 68\text{m} \tag{7.18}$$

$$C_Z = 8^2/(2 \times 4, 246)$$
$$= 75\text{m} \tag{7.19}$$

Onto these corrections we need to add 0.6 of the first Fresnel Zone to avoid diffraction losses:

$$FZC = 0.6(\lambda \cdot d_1 \cdot d_2/d)^{0.5} \tag{7.20}$$

Where d_1 is the distance of the feature from A, d_2 is its distance from B, and d is the total path length (all parameters in meters).

So:

$$FZ_X = 0.6(0.05 \cdot 14000 \cdot 36000/50000)^{0.5}$$
$$= 13.5\text{m} \tag{7.21}$$

$$FZ_Y = 0.6(0.05 \cdot 24000 \cdot 26000/50000)^{0.5}$$
$$= 15\text{m} \tag{7.22}$$

$$FZ_Z = 0.6(0.05 \cdot 42000 \cdot 8000/50000)^{0.5}$$
$$= 11\text{m} \tag{7.23}$$

Our correction table (all dimensions in meters) is shown in Table 7.2.

Entering the total corrections onto the path profile (Fig. 7.9), we see that feature Y is dominant, and drawing a straight line from antenna A through the clearance mark at Y we find that the antenna at B needs to be 140m above ground level.

We can now draw up the link budget:

TABLE 7.2
Correction Table

Feature	*Distance from A*	*Distance from B*	C	*FZC*	*Total*
X	14,000	36,000	23	13.5	36.5
Y	24,000	26,000	68	15	83
Z	42,000	8,000	7.5	11	18.5

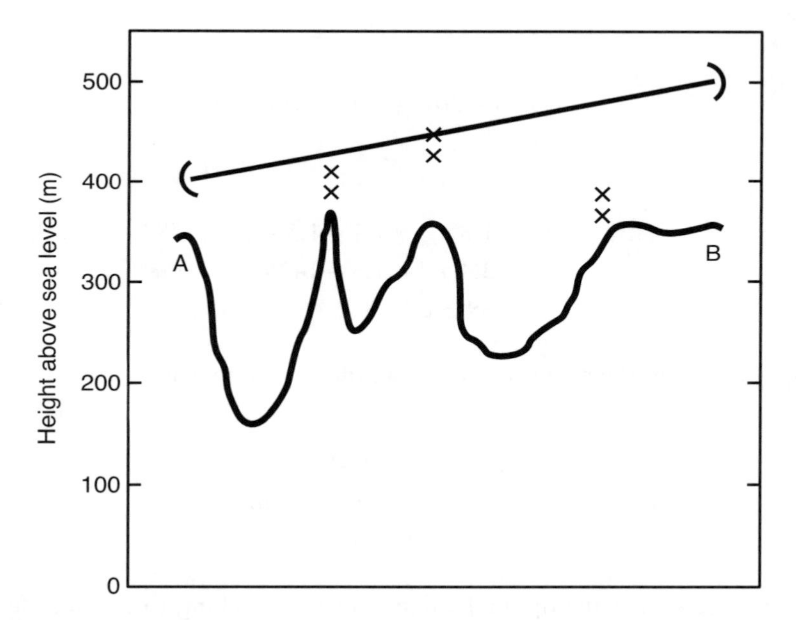

FIGURE 7.9 Determination of antenna height at terminal B in the planning exercise.

$$L = L_{FT} + L_M + L_{DM} + L_{FS} - G_A \tag{7.24}$$

where

L is the total loss between Tx and Rx.

L_{FT} is the total feeder loss.

L_M is the multiplexer loss.

L_{DM} is the demultiplexer loss.

L_{FS} is the FS loss.

G_A is the combined gains of the antennas.

Now the FS loss $L_{FS} = 32.5 + 20\log d + 20\log F$
where d is the path length in kilometers, and F is the carrier frequency in megahertz.
So:

$$\begin{aligned} L_{FS} &= 32.5 + 20\log\ 50 + 20\log\ 6{,}000 \\ &= 142\ \text{dB} \end{aligned} \tag{7.25}$$

$$\begin{aligned} \text{Thus } L &= 1.8 \times 1.9 + 1 + 1.5 + 142 - 78 \\ &= 3.4 + 1 + 1.5 + 142 - 78 \\ &= 69.9\ \text{dB} \end{aligned} \tag{7.26}$$

The input power to the receiver under nonfaded conditions is

$$\begin{aligned} P_{Rx} &= Tx\ \text{power}\ - L \\ &= +30\ - 69.9\ \text{dBm} \\ &= -39.9\ \text{dBm} \end{aligned} \tag{7.27}$$

This is less than the optimal value, and by switching to low-loss feeders we gain 2.1 dB and hence achieve the target input power to the receiver of –38 dBm.

Note that with the exception of estimating system outage (below), the hop-planning technique given is applicable to both analog and digital systems.

The only thing we now need to know is whether we require to use diversity on the link.

The United Kingdom model for multipath fading is

$$F_{0.1} = -28 + 35\log d + 8.5\log f - 14\log R + G \tag{7.28}$$

where

$F_{0.1}$ is the fading exceeded for 0.1% of the worst-month.

d is the hop length in kilometers.

f is the frequency in gigahertz.

R is the ground roughness in millirads.

G is the geographical factor.

Using the given data:

$$\begin{aligned} F_{0.1} &= -28 + 35\log 50 + 8.5\log 6 - 14\log 10 + 4 \\ &= -28 + 59.46 + 6.6 - 14 + 4 \\ &= 29 \text{ dB} \end{aligned} \tag{7.29}$$

For a nondiversity situation, the slope of the fading characteristic is 10 dB per decade. Plotting this characteristic—(Figure 7.10)—starting at the $F_{0.1}$ value, we find that the flat-fade margin of the receiver (44 dB) is exceeded for $2.5 \cdot 10^{-3}$% of the worst month and diversity is required.

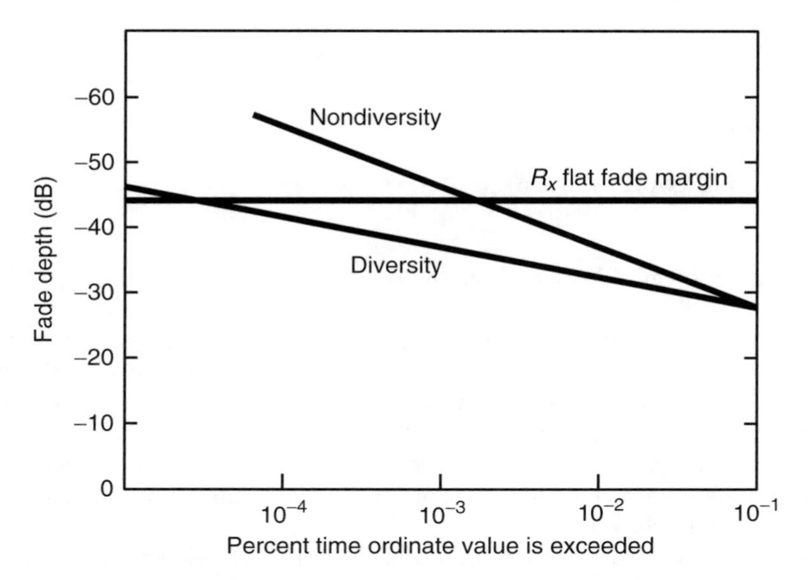

FIGURE 7.10 Cumulative distributions of fading for diversity and nondiversity situations in the planning exercise.

For a diversity situation, the slope of the fading characteristic is 5 dB per decade and plotting this, again starting at the $F_{0.1}$ value, we find the flat-fade margin is exceeded for 7.10^{-5}% of the worst month, realizing the target performance.

The minimal separation of main and diversity antennas must be in the range 150 to 200λ. At 6 GHz λ = 5 cm, the minimal separation must be 7.5m.

The diversity antenna is normally placed below the main antenna so at A, its height is 42.5m above ground level, and at B it is 132.5m above ground level.

An alternative approach to determining the reduction in system outage when using space diversity, especially if for some reason the diversity antenna differs in size from the main antenna, is to use the diversity improvement factor [2], reproduced below:

$$I = 0.0012 \cdot S^2(f/d) \cdot 10^{(F-V)/10} \tag{7.30}$$

where

I = improvement factor.

F = fade depth (decibels).

S = vertical separation of antennas center-to-center (5–15m).

V = difference of the two antenna gains (decibels).

f = frequency (2–11 GHz).

d = path length (24–70 km).

7.3.3 The Problem of Water Along the Route

If a link crosses an expanse of water or marshy ground along its path, a destructive multipath situation is possible. A link that crosses still water or marshy ground could cause a mean depression of the signal, due to a stationary multipath notch close to or within the system bandwidth. If it crosses tidal water then the signal suffers a cyclic fading pattern as the interference pattern moves from a destructive situation to additive, with the system geometry slowly changing as the height of the reflection point slowly varies.

Note that if the water is tidal, then all calculations should be carried out for the peak height of the spring tides.

A simple set of rules will establish whether there is a problem:

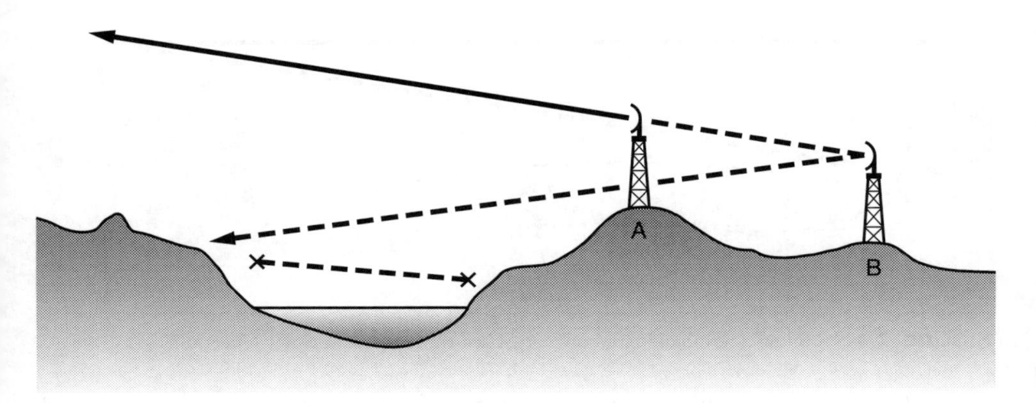

FIGURE 7.11 Moving an antenna behind high ground to avoid the problem of water along a route.

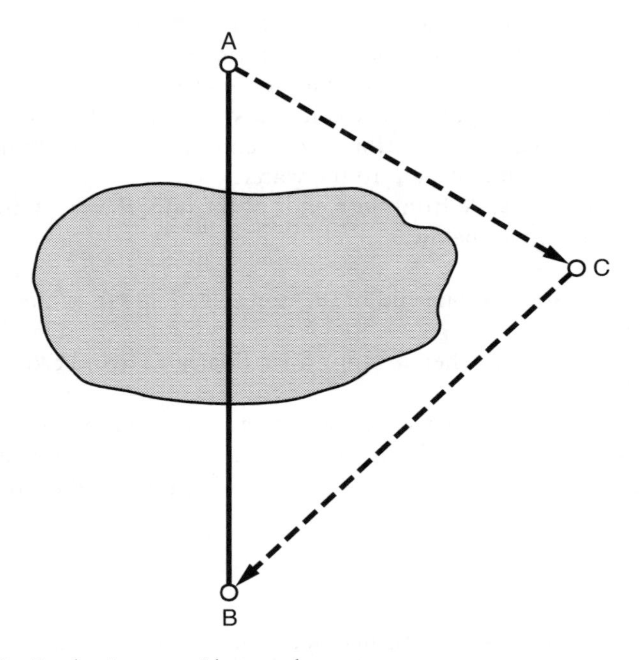

FIGURE 7.12 Doglegging to avoid water along a route.

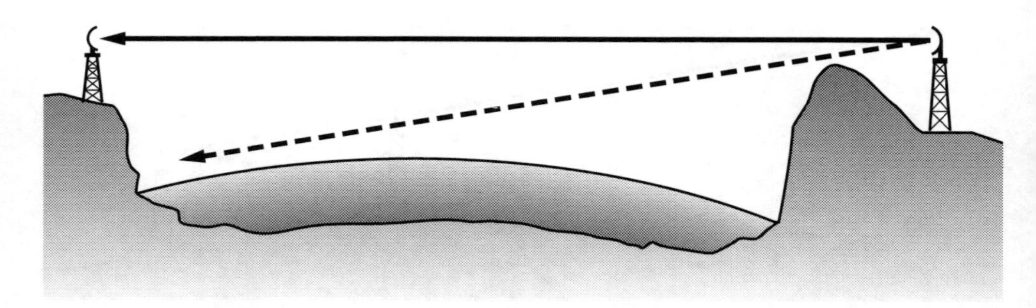

Figure 7.13 Siting a terminal inland on a cross-sea route to avoid water reflection.

1. Calculate Earth's curvature correction for the major features on the path profile and the water, using a k factor of 4/3. (4/3 rather than 2/3 since the water is always present and thus we need to use the median value.)
2. Establish whether the water is obscured from the higher antenna at both ends of the hop. If this is the case then there is no problem.
3. If the water is visible from one end of the link only, then this is a fall-back position. It is acceptable if it is not easy to reach the situation in step 2 above by moving the location of one end of the hop so that high ground blocks out the line of sight to the water.
4. If the water is visible from both ends of the hop, then you must adopt one of the alternatives below.

- Move the location of one end of the hop (Fig. 7.11) to achieve the fall-back position.
- If this is not feasible, then arrange for a dogleg to avoid crossing the water. (Fig. 7.12)
- If the hop crosses a stretch of sea, then siting the antenna inland often allows the edge of a high cliff to provide the necessary screening of one antenna from the sea under $k = 4/3$ and spring-tide conditions (Fig. 7.13).

7.4 SUMMARY

When determining that a link being planned will achieve the best possible performance, ensure that the receivers have the optimal input power under clear-sky conditions. Thus you need to know what loss to expect between the output port of the transmitter and the input port of the receiver. The calculation of this (the *link budget*) involves the losses and gains in the various parts of the hardware involved, together with what is known as the *free space transmission loss*. This term relates to the transmission loss experienced between two isotropic antennas operating within a perfect and unlimited environment with no outside influence.

If a hop is well designed, then this is the expected loss between the antennas at each end of the hop. If however, the design is at fault and there is insufficient Fresnel-zone clearance between the radio path and the terrain being crossed, under the most subrefractive conditions likely to be met, then there are times when an additional loss from diffraction is experienced. Link planning in the United Kingdom is carried out using a k value of 2/3 and 0.6 first Fresnel-zone clearance, although there are alternative rules in other countries.

The other components of the link budget are the losses in the multiplexers and demultiplexers, the losses in the feeders, and the realizable antenna gains. The availability of low-loss feeders permits the planner to fine-tune the link budget if the input to the receivers is below optimum, by a small amount, when using standard system components.

Multihop links are engineered to avoid interference at any of the repeaters or terminals by dog-legging successive hops and switching the channel polarization and the half-band used at each repeater. Also, alternate channels are transmitted from the main and diversity antennas, which simplifies the multiplexer design and allows engineering work on an antenna without losing the whole system traffic capacity.

Hop-planning is a simple procedure in which the heights of the antennas at the two ends of the hop are determined first, by correcting the salient features drawn on a flat-Earth basis for $k = 2/3$ conditions and then making a further correction to ensure that the correct Fresnel-zone clearance is available. A flat ray can then be drawn such that it does not lie below the corrected height of any of the features.

If there is water anywhere along the path, examine the situation under $k = 4/3$ conditions to see whether the water is invisible to both antennas, in which case there is no problem. If the water is visible to just one of the antennas, then this fall-back situation is acceptable, if it is not possible to make a simple change to the path geometry and attain the optimal condition. If the water is visible to both antennas, then at least one antenna location must be changed to avoid severe multipath problems.

References

[1] Vigants, A., "Microwave Radio Obstruction Fading," *BSTJ*, Vol. 60, No. 8, August 1981, pp. 785–801.

[2] Vigants, A., "Space-Diversity Engineering," *BSTJ*, Vol. 54, No. 1, January 1975, pp. 103–142.

PART II

▼▼▼

Mobile Systems

CHAPTER 8

Basic Mobile Propagation Topics

8.1 INTRODUCTION

This chapter, first in the mobile section of the book, takes the form of a basic introduction to propagation over a wide area of the radio spectrum, ranging from the very-high-frequency (VHF) band, also known as the *metric band,* covering the 30- to 300-MHz range, to the extremely-high-frequency (EHF) band, alternatively known as the *millimetric band,* which covers the 30- to 300-GHz range. This chapter also introduces a wide, but by no means exhaustive, range of "mobile" services.

The definition of a mobile communications link that we use in this section is "that the circuit established is not hard-wired into the public switched telephone network (PSTN) along the whole of its length." Thus a link established between two people, one using a "wired" telephone and the other using a car telephone, is classed as a mobile, as would a connection between two people using hand-held mobile telephones. This definition also covers a unidirectional transmission device such as a pager.

As we will discover, the term *mobile communication* is not confined to speech channels, but because of the basic nature of this book, the emphasis is, in general, directed toward speech links.

8.2 THE DIFFERENT TYPES OF MOBILE SERVICES

Envision a very basic mobile service operating between low-power, hand-held units. However, if you look at the user's environment for hand-held units, the likelihood that a good communications circuit will be established is fairly remote. If both users are at street level, then the radio path suffers diffraction losses at each end of the link from the presence of buildings in the intervening path. Establishing communications between the two mobiles using a base station, with a high antenna clear of local rooftop level, guarantees a much more reliable circuit. Further, routing calls through a base station gives the mobile access to both the PSTN and, through interbase station links, to mobiles in more remote locations.

Thus the main interest to engineers is to establish good quality communications between a base station and a vehicle or hand-held mobile. Remember that cordless communications within an office also falls within the realm of mobile communications.

Let us take a look at a number of services that you will encounter, their basic characteristics, and current importance.

8.2.1 Wide-Area Radio-Paging

The radio-pager has a receive-only capability and alerts the user, by a beeper or vibrator, to someone's attempt to contact him. Thus it is very limited in information capacity, but it is a flourishing service because it enables basic contact with someone on the move. One current drawback, however, is that, at present, the person initiating the paging call receives no confirmation that the call has been received.

8.2.2 Display Paging

This provides a similar system to the above except that it can display a message of around 80 characters long. Thus the user can receive basic information concerning the reason for the call, a telephone number to ring, and so forth.

Low frequencies can diffract round buildings, for example, quite well, whereas this is not the case for the radio-paging systems, which operate in the 80- to 960-MHz region of the radio spectrum. Thus radio shadowing on the nonilluminated side of buildings and hills is possible. These frequencies reflect well off most hard surfaces, however, and these reflections generally tend to infill the radio shadows.

The frequencies in use have certain other advantages. They penetrate most buildings well, and the dimensions of suitable antennas are such that they can be integrated into the pagers.

The lowest frequencies were the first allocated to the paging services, but as more frequencies were required, paging had to fit in with numerous other radio systems also seeking spectrum allocations. Thus the slots allocated had not only to be within bands designated for the mobile services, they had to fit between frequencies already earmarked—hence the movement to higher bands. At the same time, however, there was good news in the fact that the development of hardware was more than keeping pace with this continual upward movement in frequency, resulting in more sophisticated pagers.

8.2.3 Cordless Telephone (CT1)

This is a truly mobile option, within the limited range of its base station (a few tens of meters), since calls over the PSTN can be received by or initiated from the cordless handset. It is cheap and reliable but only has one mobile per base station, with a very limited number of channels. It operates in an analog mode with no privacy.

8.2.4 Cordless Telephone (CT2)

This was developed to overcome some of the limitations of the basic cordless telephone. It uses digital speech, which permits encryption, and has many more channels, dynamically allocated in the 860-MHz band. The transmitters are low-power (10 mW), which gives a maximum range of around 100m. The same frequency is used for transmission and reception using an interleaved mode—1 ms of transmission followed by 1 ms of reception—giving the user an illusion of a continuous link.

When the user wants to make a call, the radio scans through the 40 allocated channels and chooses the one with the least interference. It then sends a paging signal that is recognized by the base station, and the call proceeds as normal. If interference becomes a problem during the call, then a better channel is automatically identified and the call reestablished. This system ensures the best use of the available spectrum to support a relatively high density of users.

8.2.5 Digital European Cordless Telephone (DECT)

This is a standard throughout Europe. It operates in a hybrid frequency-time division multiple-access structure (10 carriers each having 12 timeslots) with time-division duplex working (both transmission directions are on the same frequency but in different timeslots, which gives the effect of continuous operation).

DECT can support Telepoint and cordless office services, and, as can be seen, has a higher capacity than CT2. It also has a higher peak output power of 250 mW compared with 10 mW for CT2, allowing the user more movement.

The allocation of the channel and timeslot, in which the mobile is working, is under the handset's control depending on the environment in which it is operating, making it a very adaptable system.

8.2.6 Telepoint

The basic concept of Telepoint, which was based on CT2 technology, is that the user has a handset to access a public base station as well as the fixed station in the user's from home. The handset has a unique identity so that the service suppliers, by the public base stations, can bill the user (and confirm that the call was made by an authorized user). The base stations were located in convenient public places such as large stores, railway stations, and town squares. The concept differed from a cellular system in that there was no mechanism for initiating a call to the mobile as the network did not know the location of a mobile at any time. Also, because there was no handover mechanism, mobile use was limited to the immediate locality of the base station.

Although the system was popular in some countries, in the United Kingdom it was very short-lived.

8.2.7 Private Mobile Radio (PMR)

This is a very useful and much used VHF mobile option used by taxis and delivery vehicles or around a farm or large industrial site. In addition, of course, emergency services and public utilities use extensive PMR systems.

8.2.8 Basic (Noncellular) Car-Phone System

This is the original 160-MHz system (in the United Kingdom) with no handover option and limited capacity. It is still in use, although it is not clear how much longer this will be the case. It suits customers who work mainly within one particular service area.

8.2.9 Analog Cellular

There is no doubt that this service was the starting point of modern mobile systems. Running at 900 MHz, the system has handover between cells and frequency reuse to make it spectrally efficient and to extend the system capacity enormously. Unfortunately, there is no common system, either worldwide or even within Europe. Despite its extended allocated spectrum, there is widespread congestion within the larger conurbations.

8.2.10 Digital Cellular

Here we are talking about widely used services that have evolved comparatively recently. Perhaps one of the best known is the pan-European system known as GSM (originally Groupe Special Mobile after the CEPT subcommittee that developed the standards to which the system works; now renamed Global System for Mobile Communications). GSM operates in the 900-MHz band, and the system has been adopted for 1.8-GHz personal communications networks (PCNs) where it is known as DCS 1800. A number of countries outside Europe have now adopted the system.

In view of the limited nature of some of the services above, the remainder of this book is focused on PMR and cellular systems, with the emphasis on the latter.

In addition to the above-mentioned services are numerous others that lie outside the scope of this book. These include satellite-mobile, traffic information, guidance and control, and vehicle location. Information on these and other areas are comprehensively covered in [1].

8.3 A GENERAL DISCUSSION

There are two very different grades of mobile communications, depending on their use. In the first situation (into which PMR falls), communication is required between a base station and a mobile to pass nonurgent messages to taxis, delivery vehicles, and so forth. It is not essential that the radio path be established at all locations and times. Taking the case of a taxi system, for instance, let us assume that there are forty taxis in the network, and at the moment a client phones in for a taxi for pickup from a given location, half of the taxis are already engaged. If, due to propagation problems, six of the remainder cannot be contacted, there are still fourteen available—the nearest to the client is dispatched and no problem arises. If it is peak time and only a couple of taxis are not engaged, but neither can be contacted, then a repeat radio call a minute or so later may well find one of them responding. We can describe this as a *messaging service*, and for this type of communication the planner aims for as high a grade of service as possible within the constraints of low cost and equipment simplicity. The planner looks for as high a base station antenna as viable, relative to the area that he or she hopes to serve, for two reasons. First, this alleviates the problem of shadowing (e.g., by buildings) between the antenna and the mobile. Second, it reduces the severity and number of black spots that cause signal blocking as the mobile goes behind high ground. CCIR Reports 239 and 567, Recommendation 370, can give guidance on likely system performance.

The alternative form of service is that of the mobile connected by a base station to a second party who is either on the PSTN or is another mobile. For this service, communication must be established at any time or location and maintained throughout the length of the call. Planning these two types of service uses different concepts.

This second type of service has, in recent times, evolved into what is known as *cellular radio* and is extremely challenging from the propagation aspect. It is planned on the basis of serving comparatively small areas from each base station and reusing frequencies in nonadjacent cells (the cofrequency cell spacing being dependent on the cell-cluster size adopted by the administration, Figure 8.1). Thus a relatively small number of channels can serve a large mobile population. This does introduce conflicting requirements, however, since we need a transmitter location that adequately serves the designated cell without any black spots but at the same time does not permit spillover into nearby cofrequency cells.

Because of the unique transmission mode in which communication is normally established, without a LOS path and relying on diffraction over and multiple reflections from buildings and the like (a truly Rayleigh fading situation), equipment designers are faced with significant problems. In LOS radio systems, one uses highly directional antennas that not only have significant gain but also a high rejection of signals coming from other than the wanted direction. Mobile systems cannot use such antennas (although some sectorization can be used at the base station), resulting in many received components of the wanted signal with significant relative time delays.

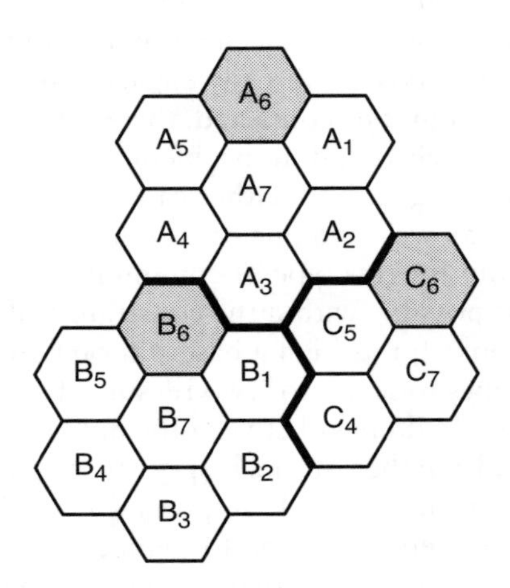

FIGURE 8.1 Cellular system—cofrequency cells in a seven-cell cluster layout.

8.4 BASIC PROPAGATION

In this section, we do not intend an indepth introduction to propagation in the mobile area but only a description of the salient features. Please note that propagation into and within buildings is covered by Chapter 11.

8.4.1 VHF Propagation

Signals above 30 MHz, unlike those in the HF band, pass straight through the ionosphere without reflection. Thus reception depends almost entirely on what is known as the *space wave,* which is one of two components of the ground wave (the other being the *surface wave*). Under the conditions present in the mobile situation, we can ignore this latter component.

In practice, even where LOS conditions exist between transmitting and receiving antennas, the FS value of signal strength is influenced by reflected components. This introduces the concept of the space wave, which in its simplest form consists of the direct and ground-reflected rays shown in Figure 8.2. More usually there are many reflected components, and the resultant vector sum at the receiving antenna

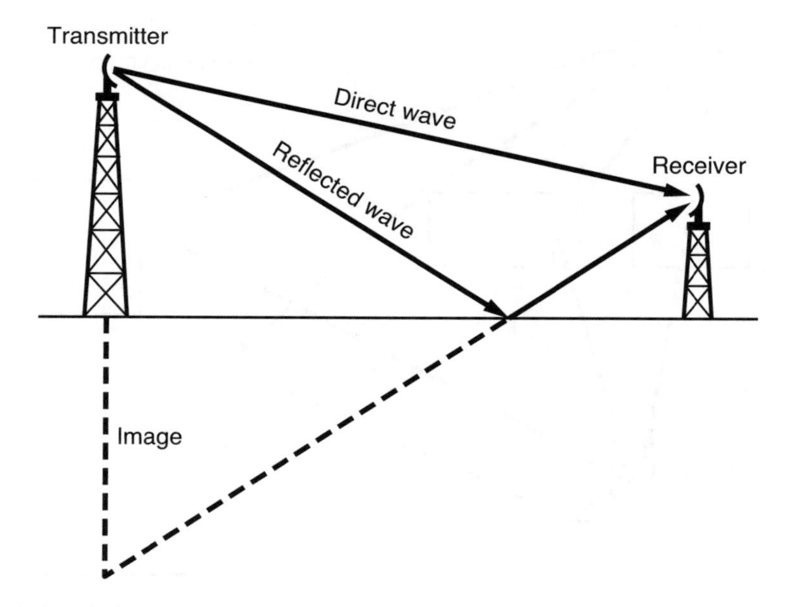

FIGURE 8.2 The basic concept of the space wave in VHF propagation.

can be very complex, (i.e., multipath transmission). In the majority of mobile situations, however, there is no LOS component, and the received signal is the vector sum of many paths resulting from reflection and diffraction situations (Fig. 8.3). The signal level can change by several tens of decibels with only a small change of the mobile's position. The rapid fading that may be experienced is the multipath term, and the much slower variation of the local mean level is a result of shadowing, for example. It is interesting to compare this situation with that found in LOS.

Fast-Fading Component

In LOS, this component is present only a small part of the time—a time-variant factor, for the most part, brought about by a two-path situation. In the mobile situation, the factor is space variant (caused by movement of the mobile under consideration or other mobiles, people, etc.) and is virtually always present, due to a large number of propagation paths.

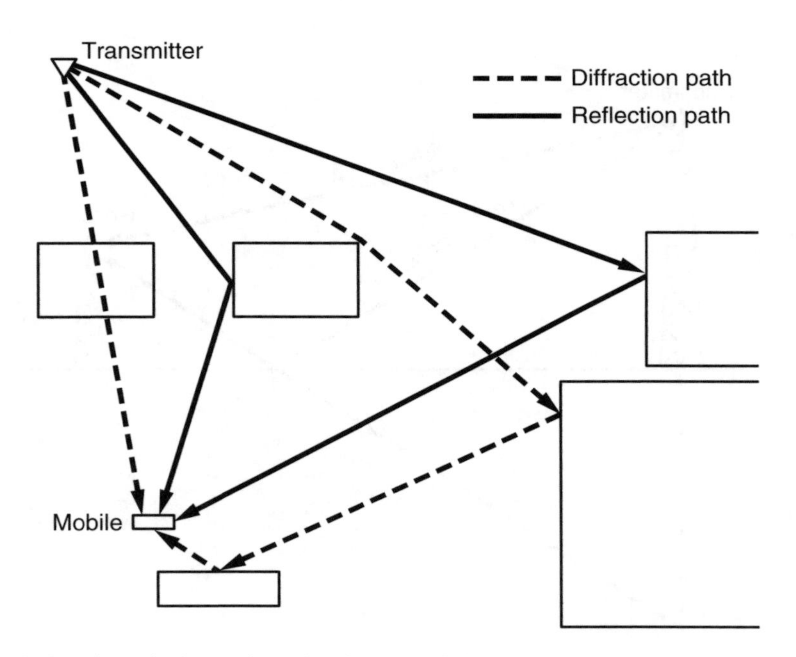

FIGURE 8.3 The multiplicity of signal paths in a mobile situation.

Flat-Fading Component

In LOS, flat fading is caused by bulk refractivity effects that lead to defocusing, offset launch, or arrival angles and again is only present for a short time. In the mobile situation, it arises from shadowing (blockage of the signal by buildings between the mobile and the base station) and is present for the majority of the time. Some of this loss, however, is counteracted by the reflection of the signals off walls, for example.

LOS Path

For a properly planned point-to-point link, the LOS path is always present, whereas in the mobile situation it is only likely to occur when the mobile is either on a straight road leading directly away from the base station or is traversing high ground.

Although in both cases the aim of the system planner is the maximum signal from one end of a link to the other, the transmission mechanisms and the planning involved are completely different, the mobile environment being quite severe.

Because of the complex nature of transmission in the mobile area, the field-strength prediction models developed by broadcasters cannot be used, and empirical methods have been developed to provide the system planner with tools. Such methods are essentially based on determining the FS path loss between transmitting and receiving antennas, adding urban loss and then adding or subtracting numerous correction factors to account for the nature of the terrain, urbanization extent, antenna height and street orientation. We discuss such methods in greater depth in Chapter 9.

8.4.2 UHF Propagation

As we move up to higher frequencies, such as to the 900-MHz and the 1.8-GHz bands, losses arising at diffraction edges increase, and the FS losses incurred are greater since there is a 20.Log F component to consider. In addition, transmission losses through walls and floors of buildings increase significantly. One compensation in the move to higher frequencies is that the spacing of diversity antennas required to achieve an effective reduction in the depth of multipath fading becomes much more practical at the mobile station. Antenna separations of as little as $\lambda/7$ result in significant diversity gain.

The effect of foliage cover has been questioned, as there might be a seasonal variation from this. However, recent measurements in rural areas have shown that the losses from tree cover are confined to those due to the trunk and branches, and the foliage itself does not contribute any measurable effect, although rain-covered foliage could add extra attenuation.

8.4.3 EHF Propagation

As mentioned in Chapter 5, one particular band that is becoming of great interest for traffic management schemes is that around 60 GHz. An initial look at Figure 5.3 (which shows specific attenuation in decibel/kilometer as a result of molecular absorption by oxygen and water vapor, plotted against frequency), suggests that the attenuation of 15 dB/km from oxygen rules out the use of the 60-GHz band. If you only require short-distance transmission, however, and wish to combine this with frequency reuse, then the band is superbly suitable for your needs. Using this band to pass traffic information to vehicles is entering the field trials area, and the small size of the antennas required, even for very narrow beamwidths, directs individual sets of information at particular traffic lanes, allowing great flexibility in the management scheme.

Although rain attenuation is significant in this part of the spectrum, the fact that links used for this service are very short means that precipitation should not create any serious problems.

In summary, we can reasonably make the following observations for the portion of the spectrum of interest:

- The higher the frequency, the greater the diffraction loss and hence the greater the shadowing effect of obstacles. The frequency range allocated to mobiles, however, reflects well off walls so that the shadow areas are often filled in by reflections from nearby objects.
- Although rain attenuation increases with frequency, the frequency allocations for the mobile services are, with the exception of the one case noted above, below that at which any untoward effects are experienced.

8.5 ANTENNA EFFECTS

8.5.1 Antenna Height

PMR and similar systems can benefit greatly from the use of a high antenna at the base station, possibly erected on a nearby hill to keep expenses down. As a rule of thumb, antenna heights have a square law effect on the received power. As with any radio system, however, you must be careful to avoid cochannel interference into other systems, so there is a limit to the benefit of this exercise.

When we consider a cellular system, we meet a conflict of requirements. We want a base station to serve its cell as well as possible, but at the same time we must avoid causing interference into other cofrequency cells. The priority is to avoid the interference—thus the base station location must be chosen carefully for optimal coverage of its cell with the lowest possible antenna height.

8.5.2 Antenna Beam Shaping

Base station antennas can have their radiation patterns shaped to obtain maximum gain in the directions in which it is required. There is obviously no point in having the sky or the ground immediately beneath the antenna highly illuminated. By stacking the antenna elements and phasing the RF input to them correctly, nearly optimal coverage of the cell is possible. Even more important, slightly tilting the array can taper off illumination toward cofrequency cells, reducing possible interference.

8.6 RADIO NOISE

Mobile radio systems are more likely to have their performance limited by cofrequency interference rather than by noise, but nevertheless, noise interference can reduce system margins. It is necessary to find out how severe this type of interference might be.

Mobile radio systems are subject to noise from a wide variety of sources. In addition to receiver front-end noise is noise from arc welders, strip lights, RF heating equipment, vehicle ignition systems, computers and other office equipment, and electrical machinery. At VHF, ignition noise is a major source of degradation to mobile radio systems [2].

There are numerous investigations of the level of man-made noise in urban areas [3–5]. Most measurements were made at street level, but base station sites have also been investigated. The general conclusion is that the level measured falls with an increase in the frequency of measurement.

A study on the suitability of the 1.8-GHz band for mobile use revealed that in that part of the radio spectrum, impulsive noise within buildings is insignificant, small handheld tools, when starting up, are the largest contributors. In an external environment, vehicle ignition noise is the main culprit, although the increase in this noise source when compared with measurements performed at 900 MHz a few years previously, was credited to increased traffic levels rather than higher individual contributions. Also, no evidence identified any particular make or type of vehicle as a primary contributor.

8.7 SUMMARY

We have identified and briefly described a number of mobile services ranging from basic paging to digital cellular. Because of the basic nature of this book, we have restricted propagation topics to the frequency bands occupied by private mobile radio and cellular systems.

We identified two types of service, the first being what was described as a messaging service, which did not require total radio coverage of the service area. Such a system is designed for as high a grade of service as possible while aiming for low cost and simplicity of equipment. The second category involves all those services in which a mobile is connected by way of a base station to a second party. For such a system, establishing communication is necessary at any time and location, maintaining it for the duration of the call.

Cellular services are spectrally efficient, using all available channels within a cluster of cells and then reusing the channels within adjacent clusters. This leads to a conflict between a base station serving a cell efficiently while at the same time avoiding spillover into other cofrequency cells.

The unique transmission mode, in which there is usually no LOS path between base station and mobile (communication being established by a multitude of diffraction and reflection paths), leads to a number of problems. Components of the received signal arrive with a wide range of delays.

As in LOS systems, there are fast-fading and slow-fading components in the received signal, but resulting from different processes. In mobile systems, the fast-fading component arises from movement of the mobile (and other mobiles, people, etc.). The slow-fading component results from shadowing by buildings between the mobile and the base station. In LOS links these components arise from the existence of atmospheric layering. The other great difference between LOS and mobiles is that fading in the former is only present for a small percentage of time, whereas in the latter, it is permanent.

Antennas in the mobile area do not have any great directional properties. The base station may use sectorization, with shaping of the radiation pattern to give maximum gain in the horizontal plane. Tilting the antenna can also help avoid cochannel interference in adjacent cell clusters.

Radio noise tends to reduce in severity as the frequency of interest increases. Vehicle ignition noise is the largest though not a serious component.

References

[1] Walker, J., *Mobile Information Systems,* Norwood, MA: Artech House, 1990.

[2] French, R. C., "Mobile Radio Data Transmission—Error Performance," *Proc. Conf. Civil Land Mobile Radio,* IERE Conf. Publ. No. 33, Teddington, 1975, pp. 93–100.

[3] Deitz, J., "Man-Made Noise," *Report to Technical Committee of the Advisory Committee for Land Mobile Services for Working Group 3,* FCC, Washington, D.C., 1966.

[4] Shepherd, N. H., "Noise Measurement and Degradation of Land Mobile Receiver Performance," *Automobile Manufacturing Association,* Detroit, 1971.

[5] Parsons, J. D., and Sheik, A.U.H., "Statistical Characterization of VHF Man-Made Radio Noise," *Radio and Electronic Engineer,* Vol. 53, 1983, pp. 99–106.

CHAPTER 9

Prediction Methods, Models, and Measurements for Mobile Systems in Rural, Suburban, and Urban Locations

9.1 INTRODUCTION

The most basic information required for any communications system is the path loss between transmitter and receiver. In the first section of this book we saw that the path loss between repeaters on a microwave LOS system can be identical to that obtained in free-space, providing that care is taken to obtain the necessary Fresnel-zone clearance under extreme subrefractive conditions.

In Chapter 8, we looked at the various mobile systems that are of particular interest and went on to discuss, in general terms, basic propagation in the bands occupied by these mobile services. We now examine the information and practical methods available to predict the path loss that we can expect between the base station and any point within the coverage area of a mobile system. We start with VHF systems and then move on to the lower microwave frequencies used in the cellular networks.

We then explore the subject of signal delays, including practical methods of delay measurement. The range of delays experienced in mobile systems is compared with those found in LOS systems.

9.2 PATH-LOSS PREDICTIONS AT VHF AND UHF

9.2.1 Rural Areas

The best known source of information for use in rural areas is that derived from broadcast services and presented as sets of curves in Report 567 of the CCIR [1]. These curves cover the 150-MHz, 450-Mhz, and 900-MHz bands. At 150 MHz, the curves are for a 3m mobile antenna height and represent path loss for 50% of the time and 50% of locations for effective base station antenna heights in the 10- to 600-m range. In using the 150-MHz curves, we can use a –3-dB height-gain factor to convert for a 1.5m antenna height. For the other two bands, the mobile antenna height is already at the more conventional value, and the effective base station antenna height is in the 30- to 1,000m range. The term *effective antenna height* refers to the height of the antenna over the average ground height between 3 and 15 km from the antenna in the direction of the mobile station. This implies that the high ground on which the base station antenna is mounted is considered part of the support structure when using these curves.

Under some circumstances, particularly for distances of only a few kilometers or if the mobile station is higher than the base station, the definition of effective height used in this chapter may lead to arithmetic results without physical significance. The following alternative definition can lead to better results on average:

$$h_1 = \begin{vmatrix} h_b + h_{ob} - h_{om} & \text{for } h_{ob} > h_{om} \\ h_b & \text{for } h_{ob} \le h_m \end{vmatrix} \quad (9.1)$$

where

h_1 is the base station effective antenna height.

h_b is the antenna height above ground level at the base station.

h_{ob} is the terrain height above sea level at the base station.

h_{om} is the terrain height above sea level at the mobile station.

The most striking feature of these curves (Fig. 9.1) is the excess path loss over the FS value. This is in direct contrast to the LOS situation, so where does this excess loss arise? The answer is that in the design of LOS systems, the planner is free to adjust the antenna heights at one or both ends of the path to obtain the necessary clearance of the radio path over terrain obstacles. It is this freedom that enables the planner to achieve FS loss between terminals, even when working under extreme subrefractive situations. The designer of a mobile system is free to adjust the height of the base station antenna, within the constraints of cochannel interference, but the

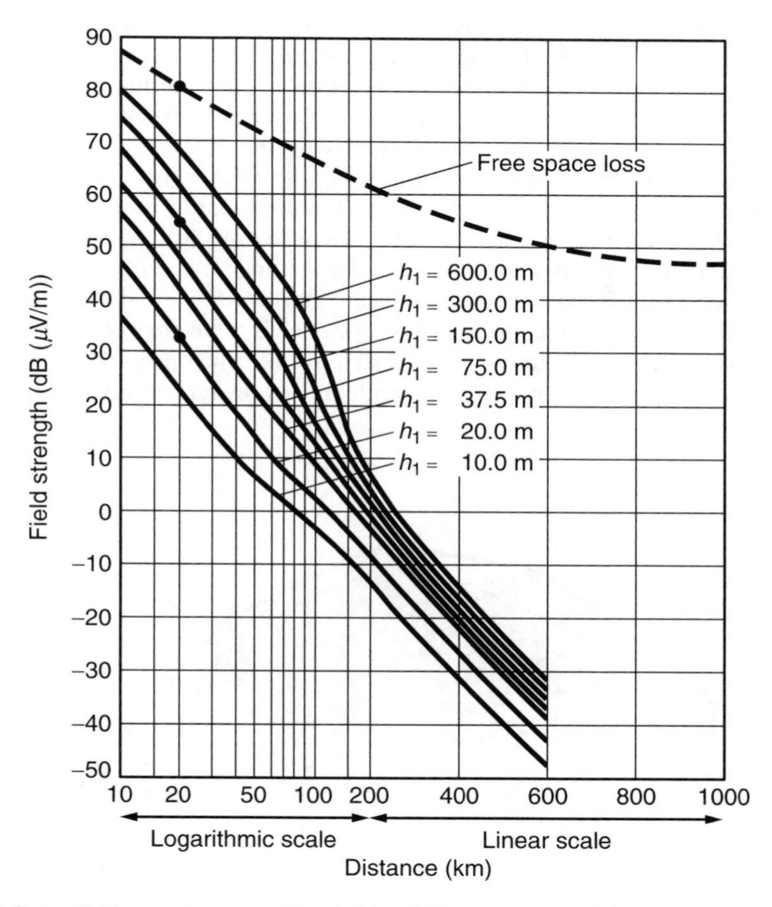

FIGURE 9.1 Field-strength curves, dB(μv/m) for 1kW erp (Reproduced from CCIR Report 567–4 by ITU permission).

mobile will, of course, be no more than a couple of meters above local ground level. Thus, signals are in the vast majority of cases, received over a diffraction path with a consequentially enhanced loss. This is the first major difference between LOS and mobile systems.

With the recent availability of detailed terrain-height databases, the opportunity to carry out diffraction path modeling has presented itself, and many organizations have developed inhouse computer programs to predict path losses introduced by multiple diffraction edges, on systems they are planning. Useful information on diffraction modeling is found in [2,3]. (COST programs are European Community collaborative studies in the areas of science and technology.)

Most methods use some means of interpreting the paths in terms of simple geometric shapes (rounded or knife edges, wedges, etc.) and apply optical theories to calculate diffraction losses. A simple example, the Deygout method [4], is shown in Figure 9.2, in which there are three predominant features between a base station and a mobile. The second of these features, producing the greatest loss A_2, divides the path into two parts. Treating the two parts as individual paths, the other two features produce losses A_1 and A_3, respectively. The total path loss is then the sum, in decibels, of the three components modified by corrections, found necessary in practice, to counteract the oversimplification of the method.

However, take care to include the effects of terrain cover in diffraction loss programs. It is no use identifying a diffraction edge at a given height above a datum level, if you ignore a belt of high trees on the edge! To avoid such problems, terrain-cover databases are being generated for use in tandem with the terrain-height information.

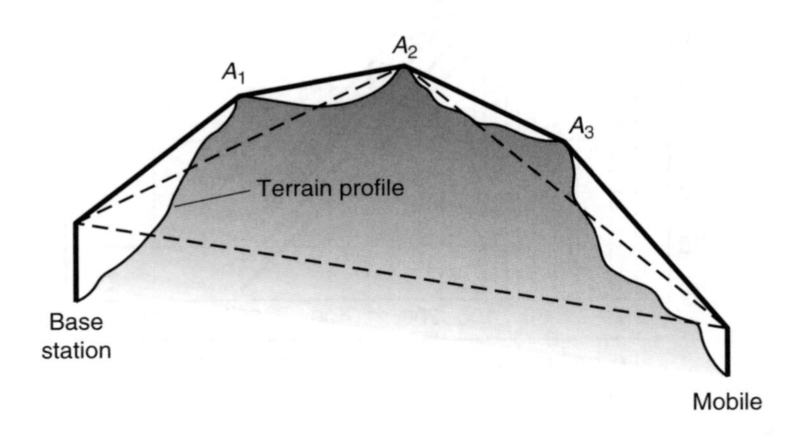

FIGURE 9.2 Calculation of diffraction losses by the Deygout method.

If a terrain-cover database is unavailable, then local surveys must be carried out to see what problems are present.

9.2.2 Urban and Suburban Areas

On moving to these areas, we can find empirical models for path-loss prediction. The best known is the Hata model [5], based on the measurements made by Okumura. The form of the model is

$$L_p = 69.55 + 26.16\log F_c - 13.82\log h_l - a(h_m) + (44.9 - 6.55\log h_l) \log R \qquad (9.2)$$

where

F_c is the carrier frequency, in the 150- to 1,500-MHz range.

h_l is the effective height of the base station antenna, in the 30- to 200m range.

h_m is the height of the mobile antenna, in the 1- to 10m range.

R is the distance between base station and mobile, between 1 and 20 km.

$a(h_m) = (1.1\log F_c - 0.7)h_m - (1.56\log F_c - 0.8)$.

We can see that although the model takes account of the above parameters, it makes no concession to the character of the intermediate terrain. The model has proved very useful over a number of years, however, and is simple to apply.

9.3 PATH-LOSS PREDICTIONS AT THE LOWER MICROWAVE FREQUENCIES

Note that the lower limit of the microwave bands is usually placed at 1 GHz. With the mobile bands at 900 Mhz and 1.8 GHz bridging this limit, however, it has become increasingly common to reduce it to 900 MHz.

9.3.1 Rural Areas

Although, as mentioned in subsection 9.1.1, there are path-loss curves for 900 MHz in Report 567 of the CCIR, their application to mobile systems is viewed with considerable suspicion. As a result, diffraction models as described earlier are increasingly being used in this area. The models provide a flexible planning tool with the option of moving the base station antenna to a number of alternative locations and observing the change in predicted coverage area, which can be displayed within a very short time interval.

9.3.2 Urban and Suburban Areas

The analog cellular networks in Europe are in the 900-MHz band. Subsequent developments have located the GSM-based digital systems in the adjacent band, and allocating more recently the PCNs to the 1.8-GHz band using the GSM-based DCS1800 system. Thus the first two systems can employ the Hata model, previously discussed, to provide a tried-and-tested path-loss model. The model as originally derived, however, has an upper frequency limit of 1,500 MHz and, as it stands, is unsuitable for use with the PCNs. COST 231 (Evolution of Land Mobile Radio (including personal) Communications) has addressed this problem and has evolved the COST 231 Hata model valid in the 1,500- to 2,000-MHz frequency range.

The COST 231 Hata model has the form

$$L_p = 46.3 + 33.9\log F_c - 13.82\log h_l - a(h_m) + (44.9 - 6.55\log h_l)\log R + C_m \qquad (9.3)$$

where

$a(h_m)$ is defined in (9.2)

$$C_m = \begin{vmatrix} \text{0dB for medium-sized city and suburban centers with moderate tree density} \\ \text{3 dB for metropolitan centers} \end{vmatrix}$$

The model is restricted to the following parameters:

F_c = 1,500 to 2,000 MHz

h_b = 30 to 200m

h_m = 1 to 10m

R = 1 to 20 km

The application of the model is restricted to situations where the base station antenna is above the rooftops of buildings adjacent to the base station.

A second COST model, the COST 231 Walfish-Ikegami model, also valid for both 900 MHz and 1.8 GHz, has evolved. Within this model [6,7] it is possible to introduce some parameters to describe the character of the urban environment, namely:

h_{Roof} is height of buildings, in meters.

w is width of roads, in meters.

b is building separation, in meters.

φ is road orientation with respect to the direct radio path, in degrees.

These parameters are defined in Figures 9.3 and 9.4. This model, however, is still statistical and not deterministic because you can only insert a characteristic value, with no consideration of topographical database of the buildings. The model is restricted to flat, urban terrain.

The COST 231 Walfish-Ikegami model is applicable to large, small, and micro cells. A steep transition of path loss occurs, however, when the base station antenna height is around the same height as local rooftops, changing slightly. Therefore apply the COST 231 Walfish-Ikegami model very carefully in this situation. For good performance of small-cell area coverage, the base station antenna should be installed several meters (e.g., more than 4m) above the maximum rooftop height of buildings within 150m.

The approaches of Walfish, Bertoni, and Ikegami are restricted by definition to radio paths that are obstructed by buildings. They are not applicable if a free LOS

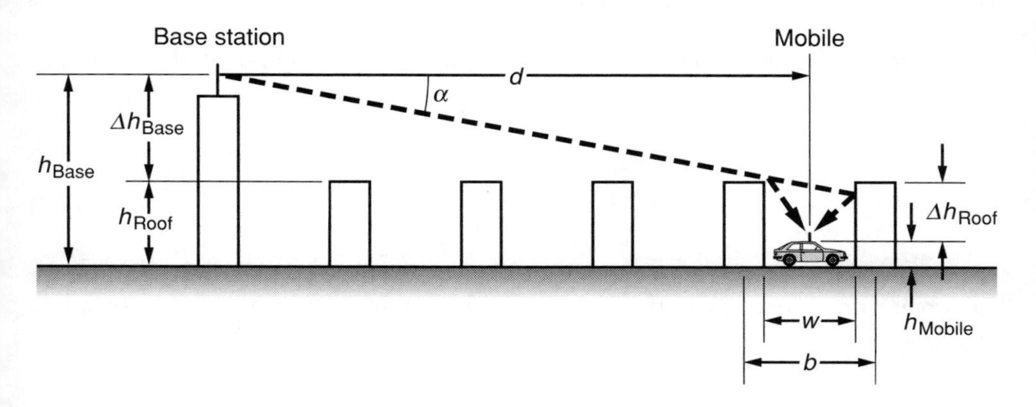

FIGURE 9.3 Parameters used in the COST 231 Walfish-Ikegami path-loss model.

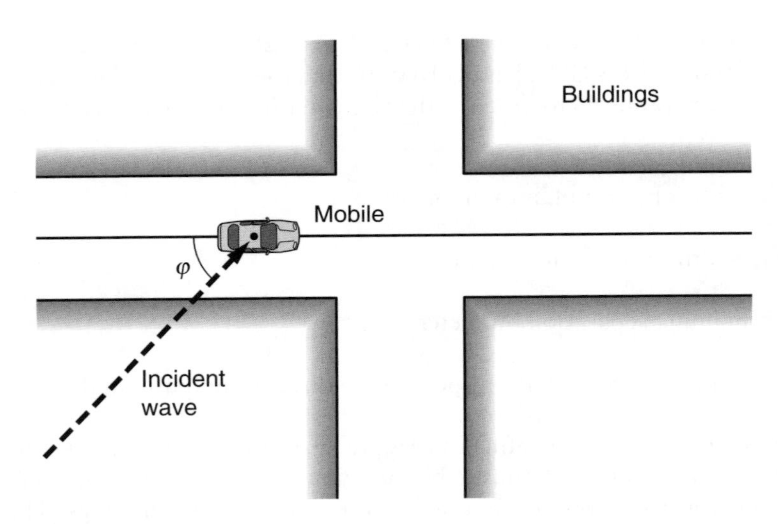

FIGURE 9.4 Definition of street orientation.

exists between base and mobile antennas within a street canyon. In a street canyon, propagation is different from FS propagation. In this case, the COST 231 Walfish-Ikegami model (described below) is based on measurements performed in the city of Stockholm [8,9]. If a free LOS exists in a street canyon then

$$L_b = 42.6 + 26 \log d + 20 \log f \text{ for } d \geq 20\text{m} \tag{9.4}$$

where the first constant is determined in such a way that L_b is equal to the FS loss for $d = 20$m.

Otherwise, the COST 231 Walfish-Ikegami model is composed of three terms:

$$L_b = \begin{vmatrix} L_0 + L_{\text{rts}} + L_{\text{msd}} \\ L_0 \text{ for } L_{\text{rts}} + L_{\text{msd}} \geq 0 \end{vmatrix} \tag{9.5}$$

The first term of (9.5) represents the FS loss L_0, the second term is the rooftop-to-street diffraction and scatter-loss L_{rts}, and the third term is the multiscreen loss L_{msd}.

The FS loss is given by

$$L_0 = 32.4 + 20 \log d + 20 \log f \tag{9.6}$$

The rooftop-to-street diffraction and scatter-loss is

$$L_{rts} = \begin{cases} -16.9 - 10 \log w + 10 \log f + 20 \log \Delta h_{Mobile} + L_{ori} \text{ for } h_{Roof} > h_{Mobile} \\ 0 \text{ for } L_{rts} < 0 \end{cases} \tag{9.7}$$

where

$$L_{ori} = \begin{cases} -10 + 0.354\varphi \quad \text{for } 0 \le \varphi < 35° \\ 2.5 + 0.075(\varphi - 35°) \text{ for } 35° \le \varphi < 55° \\ 4.0 - 0.114(\varphi - 55°) \text{ for } 55° \le \varphi \le 90° \end{cases} \tag{9.8}$$

$$\Delta h_{Mobile} = h_{Roof} - h_{Mobile} \tag{9.9}$$

$$\Delta h_{Base} = h_{Base} - h_{Roof} \tag{9.10}$$

The multiscreen diffraction loss is

$$L_{msd} = \begin{cases} L_{bsh} + k_a + k_d \log d + k_f \log f - 9 \log b \\ 0 \text{ for } L_{msd} < 0 \end{cases} \tag{9.11}$$

And the terms L_{bsh}, k_a, k_d, and k_f are as follows:

$$L_{bsh} = \begin{cases} -18 \log (1 + \Delta h_{Base}) & \text{for } h_{Base} > h_{Roof} \\ 0 & \text{for } h_{Base} \le h_{Roof} \end{cases} \tag{9.12}$$

$$k_a = \begin{cases} 54 & \text{for } h_{Base} > h_{Roof} \\ 54 - 0.8\Delta h_{Base} & \text{for } d \ge 0.5\text{km and } h_{Base} \le h_{Roof} \\ 54 - 0.8\Delta h_{Base} .d/0.5 & \text{for } d < 0.5 \text{ km and } h_{Base} \le h_{Roof} \end{cases} \tag{9.13}$$

$$k_d = \begin{cases} 18 & \text{for } h_{\text{Base}} > h_{\text{Roof}} \\ 18 - 15.\Delta h_{\text{Base}}/h_{\text{Roof}} & \text{for } h_{\text{Base}} \le h_{\text{Roof}} \end{cases} \tag{9.14}$$

$$k_f = -4 + \begin{cases} 0.7 \cdot (f/925 - 1) & \text{for medium-sized cities and suburban centers with moderate tree density} \\ 1.5 \cdot (f/925 - 1) & \text{for metropolitan centers} \end{cases} \tag{9.15}$$

The term k_a represents the increase of the path loss for base station antennas below the rooftops of adjacent buildings. The terms k_d and k_f control the dependence of the multiscreen diffraction loss versus distance and radio frequency, respectively.

If the data on the structure of buildings and roads are unknown, the default values given below are recommended.

$$\begin{aligned} b &= 20.....50\text{m} \\ w &= b/2 \\ h_{\text{Roof}} &= 3\text{m}\{\text{number of floors}\} + \text{roof} \\ \text{roof} &= 3\text{m for pitched} \\ & 0\text{m for flat} \\ \varphi &= 90° \end{aligned}$$

The COST 231 Walfish-Ikegami model is restricted to

$$\begin{aligned} f &= 800\text{–}2\text{ GHz} \\ h_{\text{Base}} &= 4\text{–}50\text{m} \\ h_{\text{Mobile}} &= 1\text{–}3\text{m} \\ d &= 0.02\text{–}5 \end{aligned}$$

9.3.3 Limits of Application

The COST 231 Hata model must not be used for $h_{\text{Base}} \le h_{\text{Roof}}$.

The COST 231 Walfish-Ikegami model has been verified for frequencies in the 900- and 1800-MHz bands and radio path lengths from approximately 100m to 3 km. In the range of base station antenna heights close to the heights of adjacent buildings, the slope of L_b versus h_{Base} is very steep.

Prediction errors are larger for

$$h_{\text{Base}} \approx h_{\text{Roof}}$$

compared with

$$h_{\text{Base}} >> h_{\text{Roof}}$$

The accuracies of prediction of the COST 231 Hata model are the same order of magnitude as that of the COST 231 Walfish-Ikegami model when using default values, but results are not equal.

The performance of the COST 231 Walfish-Ikegami model is poor for

$$h_{\text{Base}} << h$$

because the terms given in (9.13) do not consider wave-guiding in the street canyons or diffraction at corners. The parameters b, w, and φ in the COST 231 Walfish-Ikegami model are not considered in a physical, meaningful way for micro cells. Therefore, the prediction errors for micro cells may be quite large.

Note that when using the COST 231 Walfish-Ikegami model, use the effective base station height when there is a significant slope on the path.

Another interesting model has recently been developed in Japan by Sakagami and Kuboi [10], which attempts to bring in many more parameters that could have significant effects on path loss. The model, which is claimed to work for base station antennas below local roof height, is of the form

$$\begin{aligned} L_p = {} & 100 - 7.1 \log W + 0.023\pi\Phi + 1.4 \log h_s + 6.1 \log \langle H \rangle \\ & -[24.37 - 3.7(H/h_{\text{bo}})^2] \log h_b + (43.42 - 3.1 \log h_b) \log d \\ & + 20 \log f + \exp [13(\log f - 3.23)] \end{aligned} \tag{9.16}$$

where

W is the road width at receiving point (5–50m).

Φ is the angle between base station and road direction at the receiving point (0–90 degrees).

h_s is the height of building on base station side of the receiving point (5–80m).

$\langle H \rangle$ is the average height of buildings around the receiving point (5–50m).

h_b is the base station antenna height relative to the receiving point (20–100m).

h_{bo} is the base station antenna height above ground level (in meters).

H is the average height of buildings around the base station ($H \geq h_{bo}$m).

d is the distance between base station and receiving point (0.5–10 km).

f is the frequency (450 MHz–2.2 GHz).

L is the path loss referred to an 80m section of road with its center d km from the base station.

Some initial tests have compared COST 231 with the Sakagami-Kuboi models in Mannheim and Darmstadt. The first of these locations is on flat terrain, and the urban area is homogeneous in nature with buildings of around 20m tall. The second location is inhomogeneous, buildings are less high, and the terrain is not so flat.

In Mannheim all three models seem to perform well, but in Darmstadt the Hata and Sakagami-Kuboi models overestimate the path loss. Modifications have been suggested for the Sakagami- Kuboi model, and if applied, the prediction in Darmstadt is comparable with the Walfish-Ikegami model.

The general feeling is that, unless there is a particular reason for using the Walfish-Ikegami or Sakagami-Kuboi models, planners can obtain reasonable field-strength predictions by using the COST 231 Hata model. The model's simplicity makes it attractive, and prediction accuracy is probably as good as any other model when considered over a range of location types.

When comparing the multitude of parameters considered in the Sakagami-Kuboi model relative to the COST 231 Hata model, the results of the comparative tests may seem somewhat strange. This only emphasizes, however, the complexity of the transmission environment for mobile systems operating in urban situations. The signal received at the mobile results from a number of reflected and diffracted paths, and the situation is so complex that the formulation of a model that you can effectively apply to any location is not possible.

9.4 PRACTICAL MEASUREMENTS OF PATH LOSS

The purpose of making path-loss measurements is to see whether the service-area predictions, however they may be made, give a realistic picture of the situation.

For the VHF situation, the terrain covered and the service requirements are such that a few spot measurements in those areas identified as most likely to have poor reception is usually sufficient for planning requirements. Once you move into the area of cellular services, however, especially in urban environments, you must adopt a more ambitious approach to area coverage confirmation. The planners will have identified a number of possible locations for the base station, and it becomes necessary to determine which best serves the cell and at the same time does not cause interference into cochannel cells. The method that has almost universally been adopted is to equip a vehicle with field-strength-measuring equipment, together with an onboard vehicle-location system, and to carry out continuous measurements while driving around the area to be covered by the cell. For an initial survey, it suffices to drive over a series of radial routes from the base station location being evaluated, followed by a number of circumferential routes. The radial routes, which can be close to LOS in many locations, give a feel for the way in which the upper limit of signal strength is falling off as the cell boundaries are approached. The circumferential routes indicate how well the cell is illuminated in the more difficult locations.

The vehicle used for the above measurements must be typical of those used by subscribers to the service to mirror the working environment as closely as possible. Measurements must be averaged over several tens of meters to eliminate the effects of rapid fading. The antenna used should have an omnidirectional radiation pattern. The transmitter at the base station site can be narrowband.

Once the best base station location is identified, the vehicle can then be used to examine the coverage of the proposed cell more closely and to examine the possibility of interference into cochannel cells. This type of survey is both quick and cheap to perform; it can be carried out under all traffic conditions; and it accurately replicates the situation in which the users will find themselves.

9.5 DELAY MEASUREMENTS

9.5.1 Practical Measurement Techniques

In Chapter 8, we mentioned the delays that you will encounter in the mobile environment. How can you measure these?

Obviously, delays could be investigated by a radar-type approach in which a short-duration, high-power pulse is launched and the delay-power profile recorded. Such an approach, however, is hardly likely to please anyone in the locality with a system up and running. The method that has been adopted is that of *channel sounders* using a sliding correlator technique. The vehicle from which the measurement is made is equipped with a transmitter using a pseudorandom sequence as the modulation source. In the receiver an inphase and quadrature channel correlator, the start time of which is synchronized to that of the transmission, runs with a slightly slower clock rate than the transmitted sequence. The result is that the correlation is carried out over a series of timeslots representing different delays. A block schematic of such a receiver is shown in Figure 9.5.

Assume that the length of the pseudorandom bit sequence is 511 bits, that the transmitter clock rate is 10 Mbps, and the receiver clock offset is 1 kHz.

Then the resolution of the device is

$$1/\text{clock rate} = 0.1\ \mu\text{sec}$$

The maximum delay measurable is

$$\text{the sequence length} \times \text{the resolution} = 51\ \mu\text{sec}$$

The time taken to measure a complete impulse response is

$$\text{the sequence length} \times 1/\text{clock offset} = 511\ \text{msec}$$

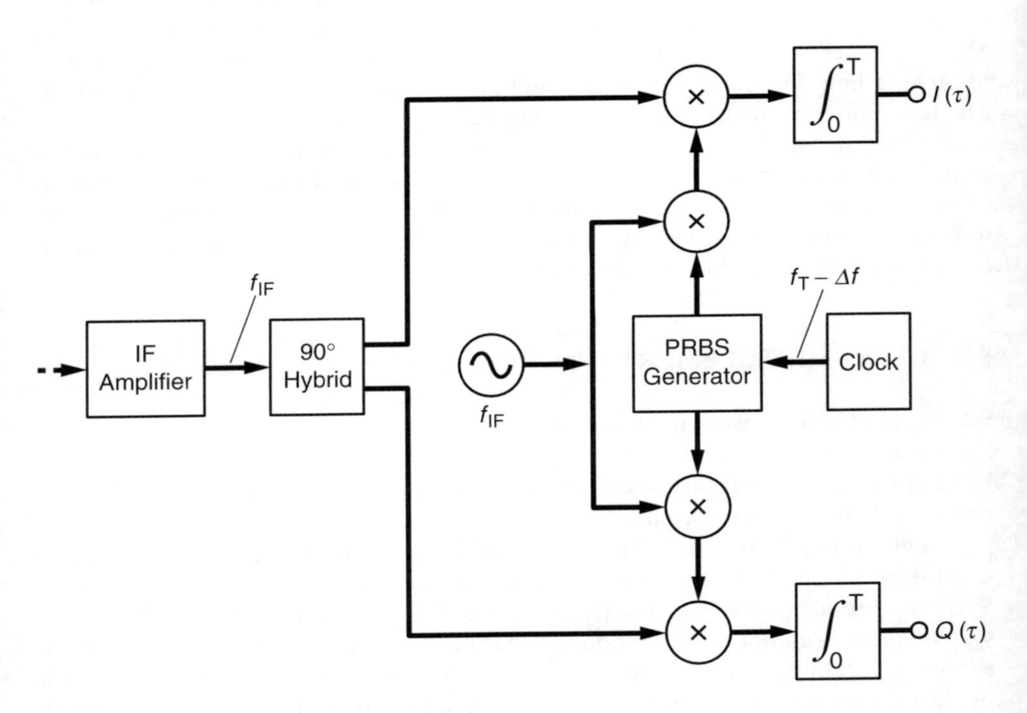

FIGURE 9.5 Principle of the swept time-delay crosscorrelator.

Thus a vehicle so equipped can drive at reasonable speeds taking phase and amplitude snapshots of the delay environment every half-second with a maximum measurable delay of 51 μsec, the upper limit of equalization in the GSM system.

It was stated in Section 2.2.1 that large-amplitude multipath signals having delays of greater than 6 ns are unlikely on LOS paths. However, we are now talking in terms of 51 μs for mobile systems, so how does this great difference arise? Consider the basic mobile scenario in Figure 9.6 in which there is a direct, although possibly diffraction-attenuated path between the base station and the mobile. In addition, there are three sources of reflected signals S_1, S_2, and S_3. S_1 lies just off the direct path, and the slight difference in path lengths introduces only a short delay. S_2 is off to one side of the base station and introduces a medium delay, and S_3, situated beyond the mobile, introduces a long delay due to the extra path length of ≃ twice the distance between the mobile and the reflector.

Now consider Figure 9.7, in which the same sources of reflection are present, but the base station and mobile are replaced by the highly directional antennas, around 1-degree beamwidth, of a LOS system. S_1 lies within the beamwidth of both antennas and hence gives rise to a multipath component at the receiver. S_2 lies well outside the main lobe of both antennas, and the reflected signal is of very low amplitude from the antenna directivities involved. S_3, while lying within the beamwidth of the transmit antenna, lies almost immediately behind that at the receiver; hence, the reflected signal is attenuated by the front-to-back ratio of the antenna used. This component is thus attenuated by something around 40 dB.

Hence, we find that the narrow beamwidths of the LOS antennas only permit

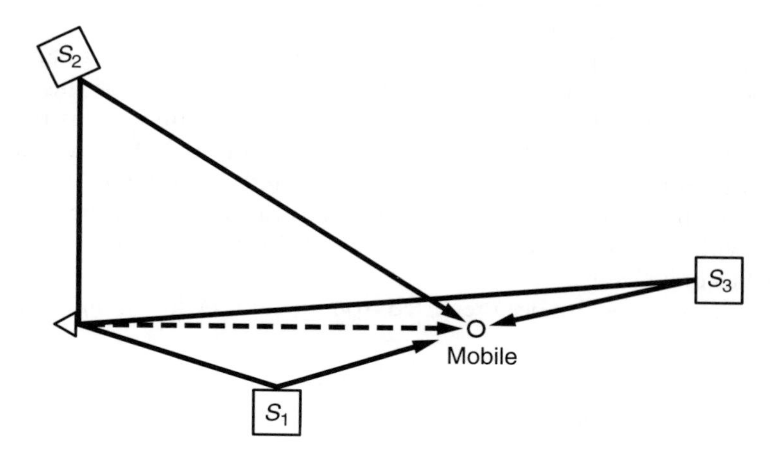

FIGURE 9.6 Multipath delays in a mobile situation.

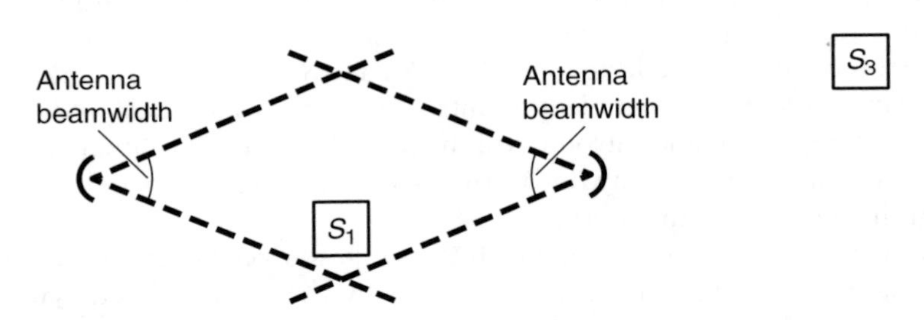

FIGURE 9.7 Reduction in the number of delayed paths when narrowbeam antennas are used

reflections from very close to the direct path between transmitter and receiver and take a major part in a multipath situation. In contrast, the omnidirectional character of the antennas used in mobile systems allows a large number of reflectors, some having very long delays, to contribute to the multipath environment. Even the use of cell sectorization does not significantly affect the situation, since all reflectors within the working sector contribute multipath signals. This is a second major difference between LOS and mobile propagation.

Can 51 μs as an upper limit of delays in mobile systems be justified? Consider an old city in Europe. It is almost certainly situated on a river, as this provided a means of transport and as such is probably one of the main reasons the city was founded there. It is also very likely that the river has, through the years, cut its way through surrounding high ground so that the city finds itself in a bowl-shaped area surrounded by hills. Thus a base station situated near one side of the city may well find itself with a LOS path to the hills on the far side of the city, with delays of around 50 μs quite likely.

Long delays have also been reported from flat areas with a range of hills in the medium distance.

9.5.2 Results Presentation

The quantities *average delay* and *delay spread* have been used in the past to present delay data statistically. These two parameters, however, are not sufficient to describe some of the important characteristics of the channel. COST 207 recognized this problem and recommended using two further parameters to describe the length of the impulse response and the distribution of energy within it. These two additional parameters are *delay window* and *delay interval*. The four parameters are defined in [11] as:

1. The average delay is the power-weighted average of the excess delays measured and is given by the first moment of the impulse response.
2. The delay spread is the power-weighted standard deviation of the excess delays and is given by the second moment of the impulse response. It measures the variability of the mean delay.
3. The delay window is the length of the middle of the power profile containing a certain percentage of the total energy found in that impulse response (Figure 9.8).
4. The delay interval is the length of the impulse response between two values of excess delay that mark the first time the amplitude of the impulse response exceeds a given threshold, and the last time it falls below it (Figure 9.9).

The importance of the last two parameters lies in their definitive character. The delay window, for instance, defines the delay range over which equalization must be carried out to recover a given percentage of the impulse. The design of equalizers is simplified if it is known that $X\%$ of the energy is contained within a given delay range, and that equalization outside of this range represents wasted effort. In simulation exercises, the 90% window ranges from 0.25 μs in rural nonhilly areas to 15.5 μS in hilly terrain, compared with the 51-μs range that GSM can handle.

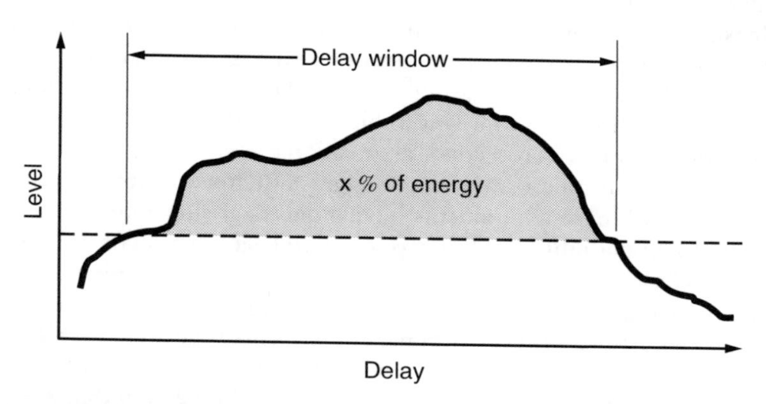

FIGURE 9.8 Definition of the term *delay window*.

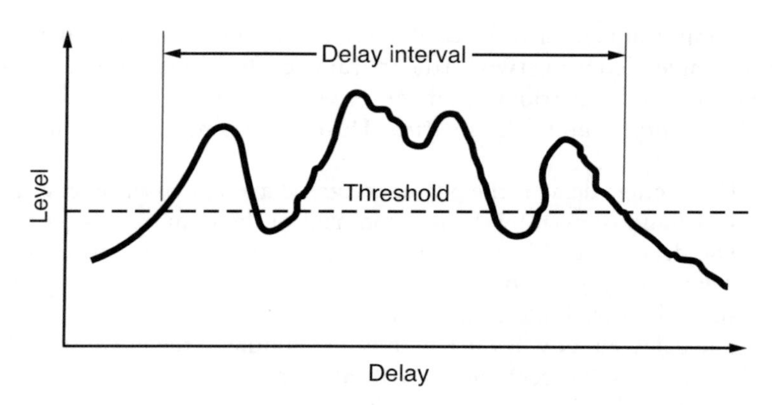

FIGURE 9.9 Definition of the term *delay interval.*

9.6 SUMMARY

Path losses in mobile systems in most cases are far greater than those experienced in point-to-point systems. This difference arises from the fact that, for LOS links, the antenna heights are chosen to give good ground clearance under the worst conditions likely, whereas in mobile situations, one end of the link is always just above ground level. Hence, a LOS path is seldom achieved, incurring diffraction losses.

We examined a number of path-loss prediction models and found that the more complex models do not necessarily produce better results. This emphasizes the very complex transmission environment involved and the difficulty of building the complexity into a model.

Measurement of field strengths in urban areas is easily carried out and provides a useful tool to determine the optimal location of a base station to serve a given area.

For digital systems, information on the delay environment is required. Techniques to measure this from a moving vehicle, in the 0- to 51-μs range, with a resolution of a fraction of a microsecond, are available.

The reason why microwave line-of-sight systems are not expected to suffer problems from multipath components having delays greater than 6 ns while mobile systems have an upper limit of 50 μs was found to be a function of the narrow-beamwidth antennas used in the former system. These antennas can only illuminate reflectors very close to the radio path due to their 1-degree beamwidth and hence the excess path loss incurred by reflected signal components is very small. However if you consider the wide coverage antennas used in mobile systems, then the base station transmitter will illuminate a large number of potential reflectors that are also visible to the mobile's antenna and thus a long delayed component of the signal arises.

References

[1] CCIR, *Reports of the CCIR, 1990, Annex to Volume 5,* ITU, Geneva, 1990, Report 567–4, pp. 310–317.

[2] CCIR, *Reports of the CCIR, 1990, Annex to Volume 5,* ITU, Geneva, 1990, Report 715–3, pp. 44–59.

[3] Commission of the European Communities, *COST 207, Digital Land Mobile Radio Communications—Final Report,* Brussels, 1989, pp. 42–45.

[4] Deygout, J., "Multiple Knife-Edge Diffraction of Microwaves," *IEEE Trans. on Antenna Propagation,* vol. AP-14, 1966, pp. 480–489.

[5] Hata, M, "Empirical Formula for Propagation Loss in Land Mobile Radio Services," *IEEE Trans on Vehicular Technology,* Vol. VT29, No. 3, August 1980, pp. 317–325.

[6] Walfish, J., and Bertoni, H. L., "A Theoretical Model of UHF Propagation in Urban Environments," *IEEE Trans on Antennas and Propagation,* Vol. AP-38, 1988, pp. 1788–1796.

[7] Ikegami, F., Yoshida, S., Takeuchi, T., and Umehira, M., "Propagation Factors Controling Mean Field Strength on Urban Streets," *IEEE Trans on Antennas and Propagation,* AP-32, 1984, pp. 822–829.

[8] Wirdemark, P., "Fitting a Two-Slope Inverse Power Law to Microcell LOS Measurements," COST 231, TD (90), 123.

[9] Berg, J. E., "Path Loss and Fading in Microcells," COST 231 TD (90) 65.

[10] Sakagami, S., and Kuboi, K., "Mobile Propagation Loss Prediction for Arbitrary Urban Environments," *Trans. Inst. Electron. Inform. Comm. Engnrs. (Japan),* B-11. Vol. J74-B-11, January 1991, pp. 17–25.

[11] CCIR, *Reports of the CCIR, 1990, Annex to Volume 5,* ITU, Geneva, 1990, Report 567–4, pp 328–332.

CHAPTER 10

PROPAGATION ASPECTS OF CELLULAR RADIO SYSTEMS

10.1 INTRODUCTION

In this chapter we introduce the development of mobile systems. This starts with the system commonly used at VHF of isolated service zones within which black spots of no communication can occur. It progresses to the basic concept of cellular systems in which "handover" to adjacent service zones provides contiguous coverage and frequency reuse both to greatly enhance the number of mobiles that can be served by a limited number of channels and to reduce the incidence of black spots. The discussion then moves on to descriptions of features required to fully develop the potential of a cellular system.

We then discuss basic features of the analog mobile system, followed by a similar look into the features of digital systems that contribute to its greatly enhanced performance compared with that of analog.

Following this explanatory material, we explore the propagation aspects of cellular systems, including a review of the techniques used to tackle the problem of increasing demands on system capacity as one moves from rural locations, through suburbia, and into city centers.

Finally, we present some definitions of the various types of cells that you will encounter, as seen from the propagation viewpoint, rather than that of traffic capacity.

10.2 THE EVOLUTION OF CELLULAR RADIO SYSTEMS

Radiocommunication is potentially the most flexible means of transferring information between any two locations. It basically requires two transceivers and their antennas; if one or both ends of the link are mobile, then radio becomes the only feasible method.

Unless the distance between the two terminals was relatively short, however, they needed access to some form of infrastructure providing suitable connections and protocols. These requirements led to the development of the early form of car-phone system operating in the VHF band. Using the United Kingdom system as an example, it comprised a number of noncontiguous service zones, each of which covered quite a large area (although with some inevitable black spots). Thus each mobile could, by way of the base station, be interconnected with another mobile, or with the PSTN.

The limitations of this scheme were, however, severe. First, only a small number of channels was available, and, as each service zone was quite extensive, only a very low density of mobiles could be serviced. Also, as a mobile approached the edge of the service zone (or a black spot), the connection became noisy, and eventually the call dropped out. Despite these limitations, the convenience of being able to communicate with others while on the move became widely recognized, and the pressure for a more flexible system with much higher capacity became very great indeed. The answer to the problem was the introduction of cellular radio.

No one appears to be credited with the idea of the cellular system. It could well have naturally developed, however, from moves to avoid black spots in the existing systems. It is clear that if the service zone were divided into a number of smaller cells, then it would be reasonably simple to site each base station for optimal coverage of its cell, eliminating black spots. Such a move, however, would introduce a multipath situation because of the different transmission times between the multiplicity of base stations and the mobile.

The next step in the development process would have been directed toward minimizing the multipath by reducing the height of the base station antennas, since each of them was now serving a reduced area. However, there would still be a multipath problem between close cells, although such a situation could have been improved by *simulcast,* a quasisynchronous transmission system. Analog signals are much more difficult to synchronize than low-rate digital ones, however. The answer to this problem would have involved re-allocating the original channels so that neighboring cells would each be served by a different set of channels, which would require some means of handing the mobile over at the interface between cells.

Thus the basic cellular concept could have evolved in this way, with all the available channels being divided within a group of neighboring cells, leaving the possibility of reusing the channels again in other, remote groups.

By a suitable choice of the frequency band, the transmitter power, and the antenna configuration, you can arrange the signal to propagate over the dimensions of its cell, with sufficient attenuation so as not to cause interference in distant cochannel cells. Thus, by carefully choosing the cell size and the number of cells in a group, frequency reuse becomes possible and the basic pattern of cells can spread without limit. The result is that a small number of channels can adequately serve a large population of mobile subscribers. The only additional requirements needed to make such a system technically viable are first, as already mentioned, a means of handing over a radio circuit from one base station to another as a mobile crosses the boundary between two cells, and second, an infrastructure linking base stations to each other and to the PSTN.

10.3 A BASIC DESCRIPTION OF ANALOG CELLULAR RADIO

The analog cellular system used in the United Kingdom is known as a total access communications system (TACS) [1], which is based on the American advanced mobile phone service (AMPS) [2]. The system details within this section are couched in general terms to give some insight into the general concepts of TACS.

10.3.1 Cell and Cluster Sizes

The basic layout of a cellular radio system is based on a hexagonal cell, with a number of such cells grouped in a cluster that uses the total number of channels available to the system. These clusters are then repeated to form an overall pattern of hexagonal cells covering any required area. The mean reuse distance (i.e., the distance between cells using the same channels) is a function of the cell size and the number of cells in a cluster. At first sight it might appear that a large cluster size has the advantage of a greater reuse distance and hence a reduced probability of cochannel interference. Looking at the number of mobiles served, however, the larger the cluster size, the smaller the number of channels per cell. Thus for a given density of users, the larger the cluster size, the smaller the cell size. This results in a greater number of cell boundaries crossed for a given distance traveled and hence a larger switching problem. Further, reducing the cell size negates the mean reuse distance advantage of the larger cluster size. For the purposes of this chapter, we will stick to a seven-cell cluster (Fig. 8.1), in which the mean reuse distance is 4.6*R* where *R* is the cell radius.

In practice, the basic hexagonal shape of a cell is not adhered to as a result of terrain features within the cell. For instance, in cells A7 and A2 in Figure 8.1, suppose there is a ridge of high ground running parallel to the boundary between these cells, situated half-way between that boundary and the center of cell A7. Now consider a mobile situated on the flank of this ridge facing cell A2. Because of the diffraction losses incurred by the signal the mobile receives from the base station in A7, the mobile may very well be better served in that location by the signal from the A2 base station. The operational standpoint considers the mobile to be in cell A2. Thus the hexagonal pattern in that area is distorted.

10.3.2 Logging On

When a mobile user switches on the mobile radio, the equipment searches through the control channels (one per base station) for the one with the highest signal level and registers with the system via that control channel by transmitting its identity code. The information that the mobile is within that cell is passed to the mobile switching center (MSC) that links base stations together and to the PSTN, in order that the system knows where to route incoming calls to the mobile. The control channel is also the initial means by which contact is made with the base station when making a call, whereupon the base station dynamically assigns an unused voice channel to the mobile. (Cellular systems are full duplex, a working method in which a call uses two frequencies, one for each direction of transmission.) The frequency bands used are in the 872- to 915-MHz range (mobile to base station) and 917- to 960-MHz range (base station to mobile). This provides 1,640-25kHz channels of 25 KHz bandwidth and includes some capacity beyond that initially provided, to overcome congestion problems, mainly in London.

10.3.3 Handover

When, during a call, the signal level from the mobile falls as it approaches the boundary between cells, the base station detects this fact and initiates the handover procedure. The MSC requests base stations adjacent to the cell occupied by the mobile to monitor the signal strength on the voice channel being used. The cell in which the base station receiving the strongest signal is situated is assumed to be the one into which the mobile has moved, and the MSC commands the mobile to retune to another channel.

10.3.4 Further Aspects

Carrier-to-Interference Ratio

It has been mentioned previously that the limiting factor in cellular system performance is the ratio of wanted signal power to cochannel interference power. The threshold ratio in TACS is around 17 dB [3].

The Effect of Multipath Transmission

In mobile systems, transmission between base station and the mobile (and vice-versa) is, as shown in Figure 8.3, by a multiplicity of paths involving diffraction and reflection with their related delays. Thus we have a typical Rayleigh fading situation. The listener will not be seriously inconvenienced by this fading activity unless the fade is very deep and system noise becomes apparent.

Diversity

Within the complex fading environment in which the mobile finds itself, the fading patterns detected on two antennas separated by as little as $\lambda/7$ is reasonably decorrelated. Thus on a vehicle-mounted system it is feasible to use switched diversity to reduce the fading effects. On a handheld device, however, the required spacing coupled to the effect of the user's body in close proximity rules out the use of diversity.

However, base station diversity is normally used and has the added benefit that the transmitted power from the mobile can be reduced while still achieving a given transmission quality. The base station is usually well removed from the clutter responsible for the multipath. Thus, to obtain the required decorrelation factor, wider spacing between the antennas is necessary.

Data Transmission

Although TACS was designed for use with speech channels, it was inevitable that eventually users would wish to pass data over the system. The provision of such a facility as an add-on to the equipment revealed several problems, not the least being those caused by multipath transmission. Although outside the scope of this book, it is interesting to note that solutions to these problems have been discovered [4,5].

10.4 DIGITAL CELLULAR SYSTEMS

10.4.1 Digital System Objectives

The 1985 installation of TACS (the first-generation cellular system) in the United Kingdom (1,000 channels) enormously increased the population of mobiles, especially as the system not only caters for vehicle-mounted units but also supports pocket-sized, handheld mobiles. Further capacity (640 channels) had to be provided later to reduce the call-blocking, especially in London, which is a good indication of the popularity of the mobile service.

Despite these additional channels, pressure on the system continued to grow. CEPT therefore put forward the concept of a fully integrated pan-European system based on GSM. The objectives were as follows:

- To define a common air interface to allow users to make and receive calls throughout Europe;
- To achieve significantly better spectral efficiency than that of the first-generation equipment;
- To be competitive in terms of functionality, performance, and cost.

Thus was born the second-generation mobile system, installed in the 890- to 960-MHz band beginning in 1991, followed by the personal communications systems—the third–generation system in the 1710- to 1880-MHz band.

In addition to increasing the potential mobile capacity, the digital systems offered the following advantages:

- Privacy by encryption;
- Integration of voice and data;
- Greater tolerance to cochannel interference.

The digital systems do not replace TACS. Analog and digital systems can operate in parallel and are able to communicate with each other in the same way, as they can contact subscribers on the PSTN.

10.4.2 Digital Speech

Digital speech on the PSTN is sampled at 8,000 samples per second (twice the maximum frequency being transmitted), each sample being of 8 bits. Thus the overall data rate is 64 Kbps. This points to an area where the capacity of the digital cellular system could be increased, if the speech could be encoded in such a way to reduce the overall data rate without too severe a degradation of speech quality. This area has inspired considerable work and not only for cellular systems. GSM uses a particular device that encodes speech at 13 Kbps and produces near *toll quality.* Acceptable encoding at around half this rate is possible in the future; the system is designed with this in mind, leading to an eventual doubling of system capacity.

This, however, is not the whole story, because these reduced bit-rate encoding schemes are susceptible to errors in the received bit stream. The techniques used for error correction introduce some redundancy, so that the gross coded bit rate transmitted increases to 22.8 Kbps (11.4 Kbps for a half-rate encoder).

GSM uses a Time-Division Multiplex (TDM) structure that allows the combination of eight signals onto each 200-KHz RF channel. A mobile transmits in one timeslot and receives in another. The modulation system is *Gaussian minimum shift keying* (GMSK), and has the following characteristics:

- Relatively narrow bandwidth, resulting in good spectral efficiency;
- A constant envelope that permits the use of simple and efficient power amplifiers;
- Low out-of-band radiation, resulting in low adjacent channel interference.

10.4.3 Data Transmission

Since transmission on the GSM system is digital, you can assume that sending data over the system presents no difficulty. Problems do arise, however, from the differing data rates and data formats used in the air interface and GSM. There are techniques to cater to these problems, and GSM can support a variety of data rates in the range of 300 bps to 9.6 Kbps.

COST 231 looked at the possibility of high-speed data transfer over the digital cellular network and concluded that this would probably need to be in the form of an overlay network operating at a frequency in excess of 10 GHz.

10.4.4 Equalization

Analog cellular systems can perform in the presence of multipath distortion, the ear being tolerant of the resultant fading and short interruptions. Digital systems, however, are not tolerant to the distortion, requiring equalization to remove this distortion to avoid system malfunction. The equalization is carried out by assessing the distortion introduced in a known bit pattern, referred to as the training sequence, and by estimating the effect on any other bit sequence. An explanation of the full operation of the equalizer is beyond the scope of this book, but it suffices to say that the equalizer is a very critical component within GSM. Without it, system performance is unacceptable.

10.4.5 Power Control

GSM uses power control at both the mobile and the base station to reduce the transmit power to the minimum required to achieve the quality objective. This is one technique that reduces the level of cochannel interference. The base station calculates the RF power level to be used by the mobile and sends an instruction to the mobile, which is capable of varying its power output from the maximum down to 20 mW in 2-dB steps.

10.4.6 Discontinuous Transmission

Another feature of GSM is the use of discontinuous transmission under the control of a voice activity detector. This is a further step in reducing cochannel interference, by around 3 dB on average.

Another advantage of this feature is the significant extension of the battery life in hand-held mobiles.

10.4.7 Handover

Unlike TACS, the GSM mobile takes an active part in the handover. The mobile is only active during two of the eight timeslots used in the TDM frame. The mobile is therefore able to scan transmissions from surrounding base stations in the other six timeslots. It reports these results, plus those from its current base station, back to the MSC (through its base station) where the handover decision is made.

10.4.8 Carrier-to-Interference Ratio

The robust character of the GSM system results in a threshold C/I ratio of around 9.5 dB compared with 17 dB for the TACS equipment. This improvement allows much closer spacing of cochannel cells without degrading system performance. It also increases total system capacity considerably.

10.5 SYSTEM CAPACITY ENGINEERING

Consider the cellular communication scenario in rural areas. There is a low density of mobiles wishing to access the system, which means that large cells can adequately serve the area. The installation of large cells results in a larger reuse distance, thus much reducing the possibility of cochannel interference.

As you move through suburbia toward the city centers the traffic capacity requirements become more demanding. Two techniques can meet this increased demand—first, the use of cell splitting in which, as the name implies, cells are split into smaller cells, each having the same number of channels as the larger cell, increasing system capacity. This technique does, however, reduce the actual reuse distance, leading to a possible cochannel interference situation unless some action is taken to avoid it. Since the size of the cell has been reduced, the transmit power can be dropped accordingly and/or the base station antenna height can be reduced.

The process can of course be repeated, but this could eventually lead to problems as the cells become very small. One such problem is the increased number of handovers required, which would stress the infrastructure. Another possibility is that the cochannel interference may become too severe. This problem, however, responds to cell sectorization involving the use of directional antennas at the base station, which splits the number of channels allocated to the cell between sectors. Figure 10.1 shows how a 120-degree sectorization can theoretically reduce the number of primary interferers from six to two. We say *theoretically,* because the reduction assumes a sharp cutoff of the sectors at *exactly* 120 degrees and also the absence of any reflectors near the edge of a sector that may "smear" the cut-off. On the same basis, 60-degree sectorization, Figure 10.2, reduces the number of primary interferers to one, making a very significant reduction in the total interfering power level.

An additional benefit from installing sectorization is the ability to tailor system capacity to demand by allocating the number of channels in any sector to the likely

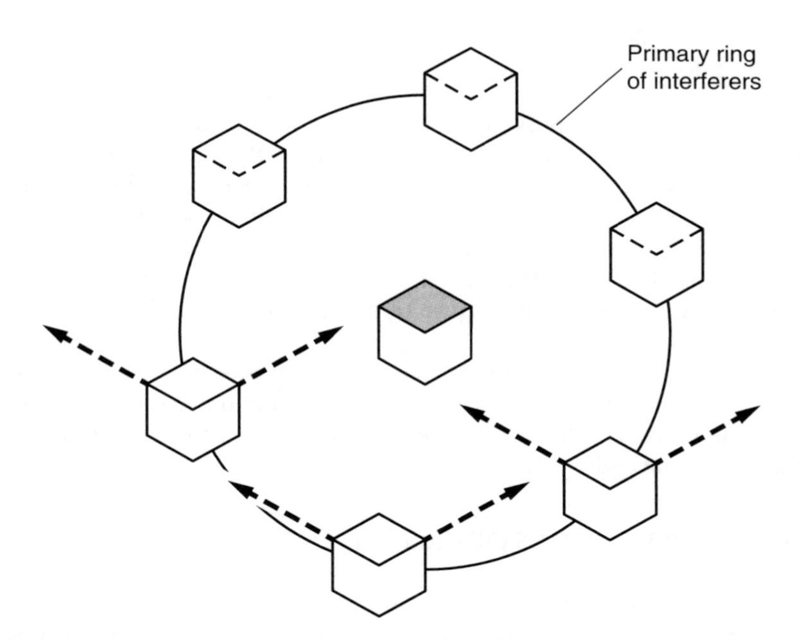

FIGURE 10.1 120-degree sectorization—two primary interferers.

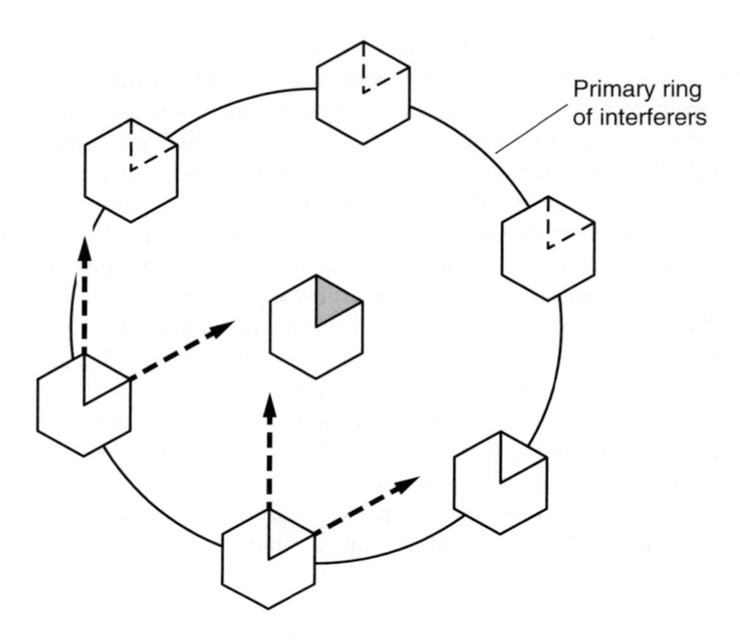

FIGURE 10.2 60-degree sectorization—one primary interferer.

number of active mobiles. A possible extension of this is the dynamic tailoring of sector capacity according to the instantaneous demand, offering more flexibility.

In certain city locations where the streets are narrow with high buildings on both sides the signal level from the base station may well be too low for reliable communication. In this situation, a microcell with a low-level antenna illuminating a short section of road may be the solution. Penetration of the signal into roads crossing the target road will be small because of the diffraction loss incurred at the corners of the intersection.

Another difficult location can be in a tunnel (or a deep cutting). For this scenario, a leaky feeder (a type of coaxial cable with a leaky outer screen so that energy is radiated from it along its length) acts, either as an independent cell or as an extension of the cell at one end of the tunnel.

10.6 COST 231 DEFINITIONS OF CELL TYPES

COST 231 realized that although there are definitions of cell types based on traffic-carrying capabilities, there was none based on cell dimensions and propagation characteristics. To rectify this situation, the following definitions were agreed.

In *large cells* and *small cells* the base station antenna is installed above rooftops. In this case, path loss is determined mainly by diffraction and scattering at rooftops in the vicinity of the mobile (i.e., the main rays propagate above the rooftops). Large and small cells differ in maximum range, which is 1 to 3 km for the latter.

In *micro cells* the base station antenna is generally mounted below rooftops. Wave propagation is determined by diffraction and scattering round buildings (i.e., the main rays propagate in street canyons in the manner of grooved waveguides). The maximum range is 0.5 to 1 km.

10.7 MISCELLANEOUS TOPICS

10.7.1 Modeling of the Transmission Path

The technique of transmitting a constant amplitude carrier and measuring the amplitude of the received signal envelope is widely used to measure path loss between elevated base stations and mobiles moving along roads [6–8]. The signal envelope is usually logarithmically compressed in the receiver so that the output voltage is in decibels.

Recordings of signal envelope variations made by many workers in the 50-MHz to 11-GHz frequency range show that because of multipath propagation in builtup areas, the envelope often follows a Rayleigh distribution when measured over an area where the mean signal strength is sensibly constant. This suggests a reasonable assumption that at any receiving point the signal comprises a number of signal components with random amplitudes and angles of arrival. The amplitudes, phases, and angles of arrival are assumed to be statistically independent.

The mean value of the signal strength changes relatively slowly as the receiver is moved. The variation is often referred to as *slow-fading* to distinguish it from the fast or Rayleigh fading caused by multipath propagation in the immediate vicinity of the receiver. The statistics of the slow fading are often log-normally distributed (i.e., the signal envelope in decibels follows a normal (Gaussian) distribution), provided the environment characteristics are reasonably homogeneous. In general, the transmission process can be modeled in three stages:

1. An inverse nth-power law relating mean received power in a given area to distance from the transmitter (e.g., $P_R = K.D^{-n}$);
2. A log-normal variation of the mean within that area, the standard deviation depending on the characteristics of the local environment;
3. A superimposed Rayleigh fading due to multipath propagation in the vicinity of the receiver.

10.7.2 Doppler Shift

The model of the received fast-fading field at a mobile is considered a large number of horizontally traveling plane waves, leading to the conclusion that the envelope is Rayleigh distributed. The Rayleigh distribution is not dependent of the spatial angle of arrival of these component waves. If you consider a mobile moving through the field with a velocity V, then each component wave is subject to a *Doppler shift* $(V/\lambda)\text{Cos}\ \alpha$, where α is the arrival angle of the component wave relative to the direction of movement of the mobile. Note that waves arriving from ahead of the mobile have a positive shift with a maximum value of $+V/\lambda$, while those arriving from behind have a negative shift. The width of the spectrum is therefore $2V/\lambda$, centered on the carrier, but to draw any conclusions about the details of the power spectral density within these limits necessitates assumptions about the spatial distribution of the arrival angles. It is common to assume that all spatial angles are equally probable, although in practice this seems unlikely.

10.7.3 Fading Rates

In the time domain, the effects of the randomly phased and Doppler-shifted multipath signals appear in the form of a fading envelope, which directly affects the performance of radio receivers. It is interesting to consider the average rate at which the envelope crosses a given level, keeping in mind that the Rayleigh fading envelope only occasionally experiences deep fades. For example, 30-dB fades occur for only 0.1% of the time.

Using a relationship derived from the work of [9,10] produces a normalized level crossing-rate diagram for a vertical monopole (Fig. 10.3), where ρ is the ratio of the specified level and the rms amplitude of the fading envelope, expressed in decibels. The maximum rate occurs at $\rho = -3$ dB. For a carrier frequency of 900 MHz and a vehicle speed of 48 km/hr, this maximum value is 39 crossing/s, falling to 3.3 when $\rho = -30$ dB.

10.7.4 Diffraction Effects Over and Around Buildings

In urban areas there is very seldom a LOS path from the base station to the mobile; the radio path is established by a mixture of reflected and diffracted components. In large and small cells, the base station antenna is above local rooftop height, and the signal propagates over rooftops, suffering scattering and diffraction loss close to the mobile. Some signal components can also suffer diffraction loss around the edges of buildings—the loss in all cases being proportional to the diffraction angle involved.

The path-loss prediction models described in the previous chapter have built-in

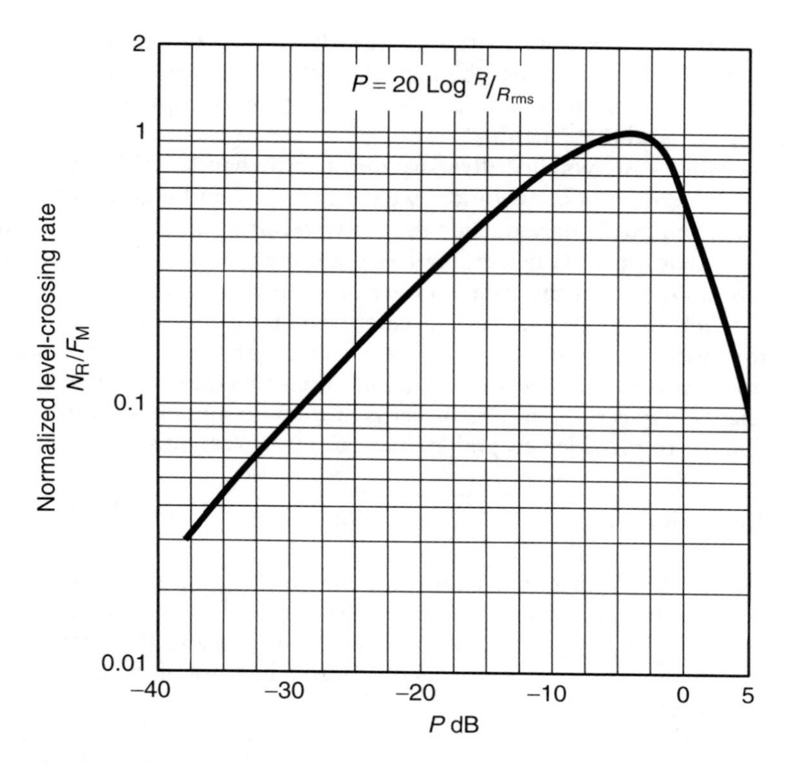

FIGURE 10.3 Normalized level crossing rates.

allowances for such losses. The Sakagami-Kuboi model identifies the height of the building on the base station side of the receiving point and the average height of buildings around the receiving point as important factors.

10.7.5 Diversity Improvement

The main purpose of diversity is to limit the signal loss from multipath fading. The required condition is that the fading patterns at the two antenna ports are reasonably decorrelated, in which case the probability that both ports will fade at the same time is small. The benefit of diversity is especially obvious when the mobile is moving slowly or is stationary, a situation in which one of the ports may be in a deep fade for a long time.

The degree of improvement in digital mobile radio performance, brought about by the use of diversity, was studied in Finland [11]. Simulation measurements demonstrated that switched diversity significantly improves the transmission of digital signals in a Rayleigh fading channel. At slow speeds, the bit-error ratio was shown to improve by up to a decade, but at higher speeds impulse noise caused by the switching reduced the benefit.

The feasibility of switched diversity in a narrowband TDMA system was reported in [12]. Here the system worked well at both low and high vehicle speeds because the switching took place outside the active receive timeslot of the equipment.

Antenna spacings of considerably less than one wavelength can provide good diversity action because of the large number of scatterers close to the mobile.

Base station diversity requires a greater spacing of the antennas than that required at the mobile because they are remote from the scatterers. Horizontal separations of 5 to 13 wavelengths between antennas are often quoted as typical. Correlation coefficients between the fading patterns at the two antennas of around 0.7 are usually sufficient to provide acceptable diversity improvement.

10.7.6 Multipath Delay

The multipath delays experienced can be influenced significantly by the environment. Although some long delays occur in urban areas, in the majority of cases the delays are of just a few microseconds. The delay spread and the number of paths involved are greater in urban areas. This is expected because of the larger number of buildings to scatter the signal.

As expected, long delays are generally more prevalent in hilly and mountainous environments than in urban areas. In moderately hilly areas such as northwest England, a figure of 10 μs is typical, although occasionally delays of up to 30 μs are observed. Even longer delays occur in truly mountainous areas. It is worth noting that GSM is designed to equalize delays of up to 51 μs.

10.8 SUMMARY

The basic VHF mobile radio system has a number of drawbacks:

- Incomplete coverage of the service area (black spots);
- Isolated service areas;
- Support of only a very limited number of mobiles.

Any attempts to improve the service could only lead to limited returns, and everything pointed to the need for a radically improved system. The main features that any new service would be required to provide were support for a much larger mobile population, automatic handover as the mobile moved between service areas, and the adoption of frequency reuse for spectral efficiency. The system adopted to meet these requirements was TACS, an analog system based on AMPS, a system used in the United States.

To achieve the basic requirements listed above, the system was based on a cluster of hexagonal cells that used the total number of channels allocated to the service among them. The clusters were then repeated to form an overall pattern of hexagonal cells. The reuse of the clusters meant that the allocated channels were used time and again, which increased the overall number of mobiles they could support, but, at a risk of degrading performance from cochannel interference. Thus such a system required careful planning so that the base station, serving any particular cell, adequately covered the cell but at the same time kept energy radiated beyond the cell boundary to a minimum.

A planned network of cells needs an infrastructure that can deal with such activities as determining in which cell to locate a mobile and handing over a call as a mobile moves from one cell to another. The mobile, when switched on, carries out the first of these by determining from which base station it is receiving the highest signal level and logging on to it. Base stations in the vicinity of the mobile monitoring its transmission control handover as the mobile approaches a cell boundary. The base stations then determine by the MSC which is best equipped to take over.

A very important factor in cellular systems is the threshold carrier-to-interference ratio of the system, which in TACS is around 17 dB. This determines the mean reuse distance between cochannel cells, which in turn determines overall network capacity.

The TACS, networks proved very popular, though even after extra channels were available, the system became overloaded. This triggered the introduction of GSM digital mobile systems in the 900- and 1800-MHz bands. This produced a robust system, the most important characteristic being that a common air interface throughout Europe would allow mobiles to make and receive calls anywhere within the continent. Other features were privacy, the integration of digital voice and data, and greater tolerance to cochannel interference by reason of its 9.5-dB carrier-to-interference threshold. This digital network is not a replacement for the analog system as the two systems work in parallel.

Two facets of GSM that reduce cochannel interference are, first, power control that adjusts the power of both mobile and base station to the minimum required to achieve the desired quality objective, and, second, a discontinuous transmission that only takes place while someone is speaking. This also reduces the demands on the battery of handheld equipment.

In rural areas, there is sufficient capacity available with fairly large cells. When we look at suburbia however, and even more so in urban areas, the traffic demands become greater. To meet these demands, cell splitting is used in which cells are split into smaller cells, each with the same number of channels as the larger cell. However, take care to avoid cochannel interference. A reduction of base station power and base station antenna height are two of the tools possible. The process of cell split- ting can be repeated, but eventually this leads to the need for an excessive number of handovers and increasing cochannel problems. This latter problem can be improved by using cell sectorization, which significantly reduces the number of primary interferers.

Difficult transmission situations respond to other techniques. In city areas with narrow streets surrounded by tall buildings, micro cells using antennas below rooftop level to illuminate a short section of a street is an effective solution. Tunnels are another problem area, but installing leaky feeders can overcome this to form either an independent cell or an extension of the cell at one end of the tunnel.

References

[1] Kemp, L. J. "A Technical Description of the United Kingdom TACS Cellular Radio System," *Mobile Phone Technology,* Memo 85/1, British Telecom, January 1985.

[2] Young, W. R., "Advanced Mobile Phone Service: Introduction, Background, and Objectives," *Bell System Technical Journal,* Vol. 58 No. 1, January 1979, pp. 1–14.

[3] Macdonald, V. H., "The Cellular Concept," *Bell System Technical Journal,* Vol. 58, No. 1, January 1979, pp. 15–51.

[4] Jarvis, R., "Data Services on Cellular Radio," *Mobile Communications Guide,* IBC Technical Services, London, pp. 56–90.

[5] Frazer, E. F., Harris, I., and Munday, P. J., "CDLC—A Data Transmission Standard for Cellular Radio," *J. Institut. Electron. Radio Engineers,* Vol. 57, No. 3, May–June 1987, pp. 129–133.

[6] Young, W. R., "Comparison of Mobile Radio Transmission at 150, 450, 900, and 3700 Mcs," *BSTJ,* Vol. 31, 1952, pp. 1068–1085.

[7] Okumura, Y., Ohmori, E., Kawano, T., and Fukuda, K., "Field Strength and Its Variability in VHF and UHF Land Mobile Service." *Rev. Electr. Commun. Lab.,* 1968, Vol. 16, pp. 825–873.

[8] Allsebrook, K., and Parsons, J. D., "Mobile Radio Propagation in British Cities at Frequencies in the VHF and UHF Bands," *Proc. IEEE,* 1977, Vol. 124, No. 2, pp. 95–102.

[9] Jakes, W. C., "Microwave Mobile Communications," Wiley Interscience, 1974.

[10] Institution of Electrical Engineers, *Radiowave Propagation,* Vol. 30, London, Peter Peregrinus, 1989, pp. 268–269.

[11] Narhi, T., "Experimental Results on Diversity in Digital Mobile Radio," *Second Nordic Seminar on Digital Land Mobile Radio Communication,* Stockholm, 1986.

[12] Commission of the European Communities, *COST 207, Digital Land Mobile Radio Communications—Final Report,* Brussels, 1989, pp. 135–147.

CHAPTER 11

COMMUNICATION INTO AND WITHIN BUILDINGS

11.1 INTRODUCTION

In this chapter we examine the closely linked topics of communication into buildings and communication within buildings.

It is easy to understand why a person using handheld equipment as he or she enters a building is not surprised if the circuit is slightly impaired, but this person certainly would not expect to experience call dropout. Thus we need to obtain a feel for the signal loss experienced through the outer structure of a range of building types. Also, if the user can continue the conversation after entering the building, how far can he or she go into the building before encountering problems (i.e., what are the inbuilding losses?).

Another scenario to consider is the working range of cordless communications within the building. Again, it is necessary to evaluate the inbuilding losses, possibly, but not necessarily, at a different frequency. DECT and the PCNs operate on very similar frequencies, but there are, of course, extensive cellular systems, analog and digital, operating around 900 MHz. Perhaps not so obvious is the need to evaluate the loss through the external structure of the building at this second frequency to establish the frequency reuse distance possibilities of cordless communications.

Whether you consider the accessibility of one of the cellular networks within a building, or the aspect of an inbuilding cordless system causing interference into co-frequency systems in nearby buildings, there is a commonality in the information requirements.

The topics covered in this chapter include techniques of making suitable measurements to answer some of these problems; some unexpected properties of the building materials encountered; a critical look at some of the models that exist; and an explanation of some possible measurement anomalies.

11.2 BACKGROUND

With the increasing popularity of handheld mobiles, it is more and more important to be able to quantify the coverage of an external cell inside a building situated within the cell's service area. The ability to continue telephone communication as a user enters a building is not just a matter of convenience. If the building is a private house, you could argue that it is an easy matter to pick up a hard-wired telephone and reestablish the call, but the user's perception of such an incident is that the system has let him or her down.

The subject of building type is important. It is easy to dismiss signal loss on entering a private house as trivial, but the term *building* can include railway stations, shopping centers, and so forth—situations where the user certainly expects normal service.

The flexibility of cordless communications within buildings, permitting a user to leave his or her normal location and yet remain in telephone contact, allowing the rearrangement of an office layout without recabling, and the like is very attractive. This recently led to considerable investigation into the propagation characteristics that are involved in this communication. In a large building it is possible that frequency reuse may be required to provide the necessary capacity. This points to one area of study—that of establishing the transmission loss through a range of commonly used building materials, not only to provide information on the possible service area from a given transmitter location but also to determine the possibility of an interference situation.

There are two mechanisms associated with building losses. The first mechanism concerns the loss through the outer structure, known as the *penetration loss.* The other refers to the losses between locations within the building resulting from the nature of materials used in the construction of the walls and floors, known as the *inbuilding losses.*

11.3 BUILDING PENETRATION LOSS

11.3.1 Measurements

A large number of measurements of building penetration loss have been carried out in a variety of locations. The basic objective of these exercises is to predict signal strength inside a typical building by extrapolating from measurements on similar buildings. It is therefore essential that the information be as reliable as possible and that the techniques used be consistent.

There are a number of definitions of building penetration loss, so let us examine three of the better known ones and assess their potential.

1. Building penetration loss is the difference in signal levels between that measured immediately outside the building at ground level and that immediately inside the building at the floor level of interest.

   This is somewhat lacking in precision and is difficult to see how measurements made by different people could be compared. It leaves a lot of questions of basic measurement technique unanswered, such as detailed below.

   Over what length of the outside and inside of the building do you measure? Do you take account of windows? Do you use any averaging? By what amount does the signal level change as you move into the building and away from the exterior wall?
2. Building penetration loss averages the signal strength over the inside of the building and compares this with the average signal strength immediately outside the building.

   This is considerably more precise, and the averaging used overcomes many of the uncertainties of the first definition. The only drawback is that averaging the signal strength over the inside of the building hides the wide floor-to-floor variation in signal strength.
3. Building penetration loss is the difference between the mean signal level measured right around the outside of the building at ground level and the mean level over the floor of interest.

   The last definition takes into account the criticisms leveled at the first two. It is considerably more definitive and popular with experimenters who are tending to base their work on this approach.

The results obtained from apparently similar buildings differ greatly from such factors as the variation in the amount of glass in the outside walls, the different ways in which the floors are divided, and the materials used. An extreme example of this is in the Electrical Engineering Department Block (A) at Liverpool University, where the penetration loss measured on the ground floor was 12.1 dB and on the first floor 1.6 dB. The reason is that the entrance is up steps to the first floor with large areas of glass and large open areas, whereas the ground floor has only a few small windows.

Thus minor problems encountered in attempting to unify measurement methods should not be taken too seriously, as they are insignificant when compared with the inherent building-to-building variations.

11.3.2 Height Gain

Another factor to consider when a signal from a remote base station is received within a building is that of *height gain.* The ground level close to the building is almost certain to be completely obscured from the base station by other buildings. The signal received at ground floor level therefore suffers diffraction loss over the edge of the nearest building between the receiver and the base station (Figure 11.1). If we next consider a floor higher up in the building, then although the signal is still received over a diffraction path, the diffraction angle (and the associated loss) is smaller. Thus, as you measure higher and higher up the building, the trend is toward less and less diffraction loss. This increase in signal level with height is known as the *height-gain effect* and has an average value of 1.5 dB/floor at 900 MHz and 2.0 dB/floor at 1.8 GHz.

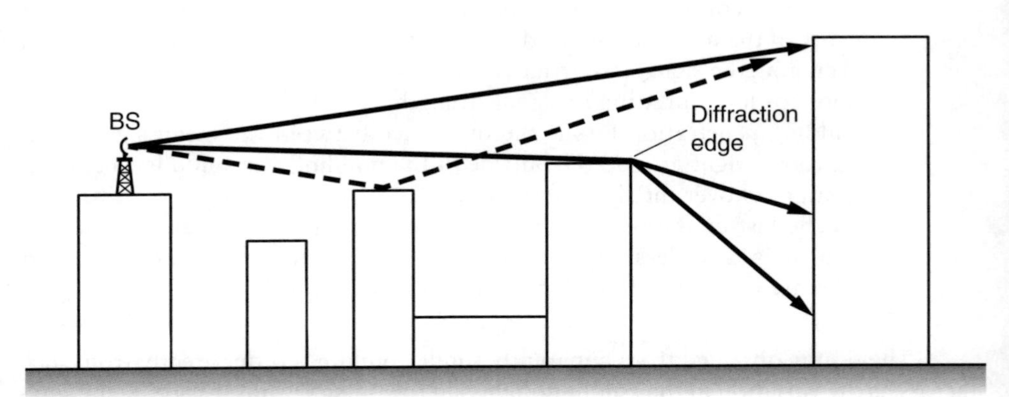

FIGURE 11.1 Height-gain effects.

11.3.3 Measurement Techniques

Measurement equipment consists of two units. The first of these, representing the base station, is a continuous wave (CW) transmitter (typically 5W) feeding a colinear antenna, mounted clear of local obstructions. The second unit represents the mobile and must be trolley-mounted for ease of use, with an antenna mounted at a height representative of a handheld device when in use (1.5m). The antenna is coupled to a signal-strength-measuring receiver with a dynamic range of 60 to 70 dB and a noise floor typically equal to –125 dBm. The output of the receiver feeds a microcomputer that handles the data sampling and processing. The sampling is performed under control of a transducer driven from one of the trolley wheels and is carried out several times per wavelength of travel. The trolley moves along a preselected route within rooms and along corridors. The data is stored on disc, and then fully processed on a separate computer to obtain the mean signal strength over one floor of a building.

The received signal statistics can be modeled as a small-scale process (multipath) with minima separated by around half a wavelength, superimposed on a large-scale process (shadowing).

11.4 INBUILDING LOSSES

11.4.1 Measurements

The measurement of inbuilding losses is straightforward. The equipment representing the base station of the cordless communications system can comprise a measuring receiver coupled to a small computer. The mobile, a small transmitter, is moved evenly around the area to be covered by the base station, with its antenna at a representative height above the floor. The receiver samples at a typical rate of 10^3 samples per second. It is worth noting that all inbuilding measurements should be carried out under conditions that are as natural as possible. It is no use carrying out a measurement program late in the evening, when a building is empty of people, just because it is more convenient. The absence of people moving around modifies the multipath and shadowing environments, and the results will not mirror real life. Body-loss measurements carried out at 900 MHz [1] showed that close to the body, at waist and shoulder height, the field strength decreased by 4 to 7 and 1 to 2 dB, respectively, compared with measurements carried out several wavelengths away.

11.4.2 Path-Loss Models

A considerable amount of work has been carried out to determine the effect of construction methods, materials, and so forth on the transmission of radio signals within buildings. We detail results of some studies in this area below.

In the United Kingdom, experiments carried out at 900 MHz, in a typical office building of several floors [2], yield a practical propagation model of the form

$$P = L + 10n \cdot \text{Log } d + kF \tag{11.1}$$

where

P = path loss (decibels)

L = path loss (decibels) at 1m from the transmitter antenna

n = distance/power law coefficient

d = slant range (meters)

k = number of floors traversed

F = attenuation per floor (decibels)

At 900 MHz, the distance/power law coefficient n was approximately 4, with L = 30 dB and F = 5.4 dB per floor.

An alternative to this expression [3], uses r equal to the horizontal range in place of d. Further work is required to determine the most suitable expression for different conditions.

A model that combines both losses within buildings and penetration losses was produced by [4,5] and is expressed in the form

$$\text{Path loss} = L + 10 \cdot n \cdot \log d + k \cdot F + p \cdot W_I + W_E \text{ (dB)} \tag{11.2}$$

where

L = the mean path loss to the building perimeter (decibels)

n = the exponent of the distance dependence

d = the distance into the building (meters)

k = the number of floors between the Tx and Rx

F = the floor-loss factor (decibels)

p = the number of interior walls between the Tx and Rx

W_I = the internal wall-loss factor (decibels)

W_E = the external wall-loss factor (decibels)

The wall-loss factors depend on the materials used in the walls. At 900 MHz these lie in the range 0.4 to 29 dB and are somewhat higher at 1,800 MHz. When considering losses through partitioning walls within a building, however, simplify the situation and consider just three classifications: *light* with a 1.5-dB loss, *medium* with 5-dB loss, and *heavy* with a 10-dB loss.

When using the model for the inbuilding situation, L becomes the path loss (decibels) at 1m from the transmitter antenna, and the term W_E is not used.

11.4.3 Measurement Results

In general, Rayleigh-distributed fast fading occurs in buildings [6–9].

The variations of the local mean values within a building can fit a log-normal distribution [7,8] and [10]. The reported standard deviations lie in the range 3 to 9, the highest values being where a LOS exists between the base station and the building.

Building penetration losses at 900 MHz fall in the 7- to 19-dB range for office buildings. Internal wall losses vary over a wide range, depending on their construction, the upper limit being around 17 dB at 1,700 MHz. Data on the dependence of building penetration losses on frequency are insufficient to make any firm conclusions. However, data on wall penetration losses indicate that the losses increase with frequency.

For diversity, measurements performed with a portable receiver at 900 MHz, inside a typical three-story office building in Italy [11], show that even for a reduced spacing ($\lambda/7$) between two quarter-wave antennas, the correlation coefficient between the two signals did not exceed 0.7 on average, and improvements larger than one decade for exceedance probability were measured at attenuation values of 20 dB or more.

British Telecom Laboratories made a study on the best way to operate an inbuilding service, linking cordless telephones into a wireless PBX [12]. The study proved that the range of the cordless telephones was restricted to around 50m due to shadowing, loss through walls, and so forth. The alternative approach was to use leaky feeders to distribute the signals around the building, resulting in handset users always being within the system's service area. This way of coupling the mobile into the PBX was further enhanced by adding amplifiers at suitable points along the feeder to make up for the cable loss.

The main conclusions were that for operating distances greater than about 20m, the leaky feeder noticeably improved the signal distribution with very even cover and much higher levels at extreme range. The improved coverage from a leaky feeder allows reduced transmitter power and protects against interference from cochannel users. The signal decay rate away from a feeder is higher than from an antenna, and so cochannel users can be packed more closely together.

11.5 COMMENTS ON AVAILABLE MODELS

The assumption made when the above models were being developed is that the dominant transmission mode is of a straight-line nature (Figure 11.2) between transmitter and receiver. This is a very sweeping assumption—the models derived, although frequently used, should be treated with caution. Once you make the assumption of straight-line transmission, wall- and floor-loss factors have to be tailored to arrive at measured path-loss figures. It is interesting to compare the resultant loss factors with those derived from direct measurements made on wall and floor samples, in an anechoic chamber.

More searching measurements recently carried out established that the lower loss transmission paths are more likely to be by way of elevator shafts, stairwells, or even through a window in the room in which the transmitter is situated. The paths are then reflected from another building and back through a window close to the receiver (Figure 11.3).

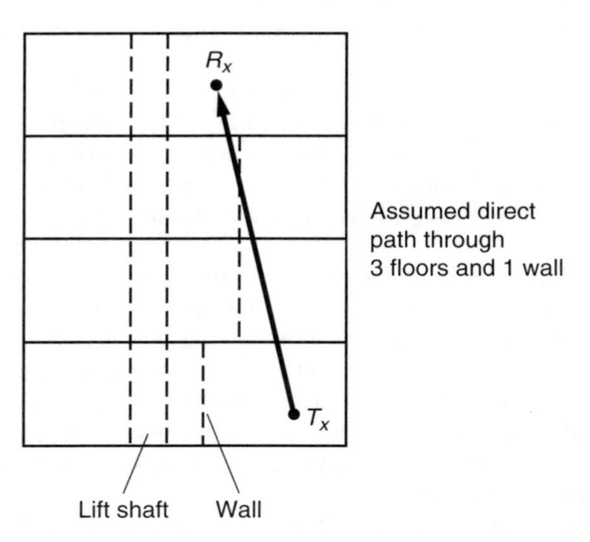

FIGURE 11.2 Assumptions made in (11.1) and (11.2).

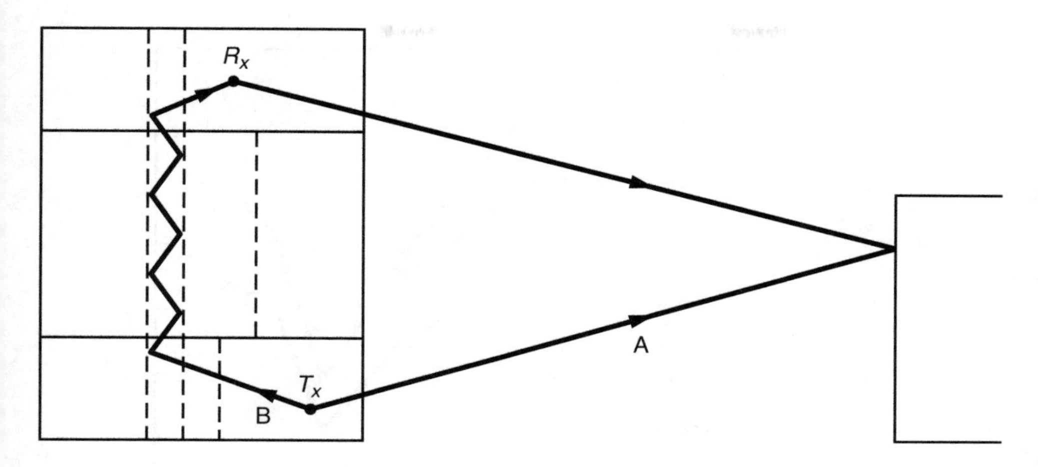

FIGURE 11.3 Alternative signal paths for inbuilding situations.

The lesson learned is that you must take great care with all inbuilding measurements and must account for any effects from alternative transmission paths if possible. This still leaves one question to be answered, however, "If you can establish the wall and floor losses, how can you predict the mean signal level in a particular location?" Since the level is, in fact, the vector sum of a number of alternative signal components, the relative phases of which vary rapidly with mobile movement. Available models do not recognize this situation. An interesting reference reporting on the direction of arrival of signals within a building is in [13].

11.6 ANOMALOUS EXPERIMENTAL RESULTS

A number of apparent anomalies have been encountered during measurement programs. We discuss three of particular interest below.

11.6.1 Penetration Loss Versus Frequency

In [7] is a table of the results of penetration loss measurements on each floor of the Electrical Engineering Department Block (A), for frequencies of 441, 896, and 1,400 MHz. The results show a very definite trend for the penetration loss to decrease with increasing frequency. A few years later, measurements were carried out at 1.8 and 2.3 GHz, and to everyone's surprise the results showed the reverse effect. The two sets of results have been plotted for three of the floors (Figure 11.4) and show a common trend of a very distinct minimum loss at around 1,400 MHz.

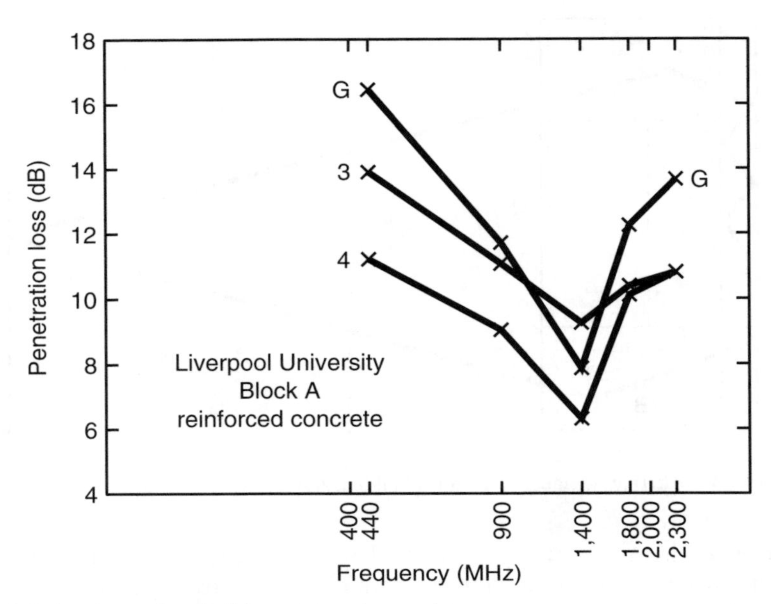

FIGURE 11.4 Anomalous building penetration results.

There are no obvious reasons as to why this should be, but the construction of the building is reinforced concrete, leading us to query whether the dimensions of the reinforcing mesh play a part. No other reports of this effect have been received.

Our point is that you must consider all available information before making any conclusions. Also, if further information becomes available at a later date, it is worth testing it against earlier conclusions to check for consistency.

11.6.2 Height-Gain Factor

Figure 11.5 is a plot of building penetration loss against floor level at 1.8 GHz. The plot shows the combined measurements from several buildings, hence the spread of results at each floor level. The general trend of the height gain factor is 2 dB/decade as you would expect at 1.8 GHz. There is an unusual factor, however, the reverse trend between floors 6 and 9. A possible explanation of this anomaly is that the relative heights and positions of the base station and the nine-story building were such that the base station had a LOS path to the top three floors. In addition, if there was a reflection off a rooftop, then the top of the tallest building could have been in the destructive part of a vertical interference pattern. Thus the reduction in the signal level from the multipath situation, shown by the interrupted lines in Figure 11.1,

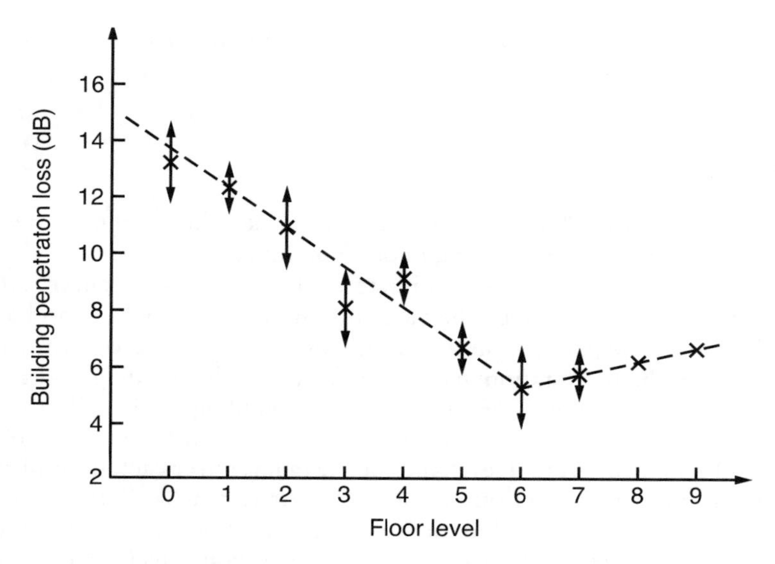

FIGURE 11.5 Unexpected results from height-gain measurements.

could have masked the height-gain factor. This scenario seems supported by the fact that when a more remote base station, with a different viewing angle to the nine-story building was used it resulted in a normal height-gain plot.

11.6.3 Wall and Window Penetration Loss

Reports have been made of out-of-character penetration losses through some windows and lightweight plasterboard partitions. The resulting investigations revealed that the windows were tinted, the tint being produced by a metallic deposit, and the plasterboard had a foil backing. The lesson learned from this is do not take building materials at their face value. It also raises the point that architects should have a good background knowledge of the needs of modern inbuilding communications techniques to design their buildings effectively.

In conclusion, all experimental measurements connected with buildings must be very carefully planned. Question your results and seek reasons for anomalies. If you have good reason to suspect that the position of the transmitter relative to the receiver is responsible for some unexpected results, then try a more suitable location for it. The remeasured results may be of a different level than before, but if the questionable trend has not changed substantially, you may have to look elsewhere for the answer to the problem.

Penetration loss usually increases with increasing frequency, but this does not always happen. The answer as to *why* is still sought.

11.7 SUMMARY

With today's popularity of handheld mobile telephones, there is a great need to predict their performance when working under inbuilding situations.

There are two particular areas of interest. The first of these concerns the excess path loss when working between the base station of an external cell and a location within a building. The path loss in such a situation comprises three terms. These are the loss between the base station and the perimeter of the building under consideration, the loss through the external walls of the building, and the loss within the building due to wall attenuation. By determining the mean signal level around the outside of a building at ground level and the mean level over each floor of the building, and then subtracting one from the other, we arrive at what is known as the penetration loss. This loss is a combination of the external wall and internal losses. By expressing it as a mean for each floor level, it gives you a feel for how successful communication using an external cell would be. The penetration loss usually decreases the higher up the building you go, because of what is known as the height-gain factor. This factor arises from the signal suffering diffraction loss over the edge of an adjacent building. The smaller the diffraction angle, the less the resultant loss. At 900 MHz, the height gain is around 1.5 dB/floor and at 1.8 GHz increases to 2.0 dB/floor.

The second area of interest is the coverage of a cordless communications system within a building. The test transmitter is set up in a suitable location and then once again the mean signal level on each floor is determined. Best coverage is often obtained with the transmitter as near the center of the building as possible, and the highest signal levels on each floor are usually close to stairwells, for example.

Measurements are made using a trolley-mounted mobile receiver with the omnidirectional antenna usually mounted at 1.5m above ground level. Samples are taken several times per wavelength of travel and are triggered by a transducer coupled to one of the trolley's wheels. The data obtained are stored in files on a computer linked to the receiver and then processed later to obtain the required information.

Several path-loss models are available but require care. A common feature is that they are based on the premise that the dominant received signal component travels on a straight-line route between transmitter and receiver, whereas researchers have identified many signal paths, some of them occasionally being low loss. It is believed that the floor and wall losses determined by using these models are not necessarily true values but come from using too simple a model in a complex situation.

Measurement programs occasionally yield anomalous results. Sometimes this is from building materials that are not all that they seem to be. Tinting glass, for instance, may be achieved by the use of a very thin metallic deposit. Plasterboard may have a foil vapor barrier on the reverse side. Both of these situations, of course, yield high-transmission loss values.

Other anomalies can arise from an unexpected propagation effect masking another in which you have an interest. Remember, the more data you have the more likely it is that you will be able to identify anomalous results.

References

[1] Akeyama, A., Tsuruhara, T., and Tanaka, Y., "920-MHz Mobile Propagation Test for Portable Telephones," *Trans. Inst. Electron. Comm. Engrs., Japan,* Vol. E65, 1982, pp. 542–543.

[2] Motley, A. J., and Martin, A. J., "Radio Coverage in Buildings," *Proc. Nat. Comm. Forum 1988,* Chicago, October 1988, Vol. 42, No. 11.

[3] Owen, F. C., and Pudney, C. D., "Narrowband In-Building Propagation for Digital Cordless Telephones at 1,700 MHz and 900 MHz," *Electronics Letters,* Vol. 25, No. 1, January 1989, pp. 52–53.

[4] Motley, A. J., and Keenan, J. M. P., "Personal Communication Radio Coverage in Buildings at 900 and 1,700 MHz," *Electronics Letters,* Vol. 24, No. 12, 1988, pp. 763–764.

[5] Keenan, J. M. P., and Motley, A. J., "Radio Coverage in Buildings," *British Telecom Technology Journal,* Vol. 8, No. 1, 1990, pp. 19–24.

[6] Cox, D. C., Murray, R. R., and Norris, A. W., "Measurements of 800-MHz Radio Transmission Into Buildings With Metallic Walls," *Bell System Tech, J,* Vol. 62, No. 9, 1983, pp. 2695–2717.

[7] Turkmani, A. M. D., Parsons, J. D., and Lewis, G. D., "Measurements of Building Penetration Loss on Radio Signals at 441, 900 and 1400 MHz," *Journal of the IERE,* Vol. 58, No. 6, (suppl), pp S169–174, 1988.

[8] Turkmani, A. M. D., Parsons, J. D., and de Toledo, A. F., "Radio Propagation into Buildings at 1.8 GHz," *COST 231,* TD (90) 117.

[9] Backman, P., Lidbrink, S., and Ljunggren, T. L., "Building Penetration Loss Measurements at 1.7 GHz in Micro Cellular environments," *COST 231,* TD (90) 121.

[10] Johnson, I. T., Johnston, W., and Kelly, F. J., "CW and Digital Propagation from Outdoor Public Cells Measured in Customer Premises Cells," *COST 231,* TD (90) 82.

[11] Damosso, E., and Lingua, B., "Antenna Diversity Improvements in Radio-Wave Propagation Within Buildings," *Proc. of the 4th Seminar on Mobile Systems,* Pontecchio Marconi (Bologna, I), October 1988.

[12] Palmer, D. A., and Motley, A. J., "Controlled Radio Coverage Within Buildings," *British Telecom Technol. Journal,* Vol. 4, No. 4, October 1986, pp. 55–58.

[13] Matthews, P. A., and Hairi Abu Bakar, A., "Direction of Arrival of Radio Signals Within Buildings," *COST 231,* TD (92) 65.

CHAPTER 12

Propagation in Fixed and Mobile Systems—A Comparison

12.1 INTRODUCTION

This book features descriptions of many facets of LOS and mobile propagation, but these are of necessity scattered through the previous eleven chapters. Because of this, it is not easy to make an overall comparison of their characteristics. The purpose of this final chapter, then, is to bring these two topics together in a single location, compare them under a number of headings, and offer explanations why, in some cases, they appear to be widely different in character.

When discussing fixed systems in this book, we placed emphasis on high-capacity microwave systems operating over long paths and meeting internationally agreed performance specifications. This is because they are ideally suited to this role, and it has become their chief use. The type of traffic carried by the systems are television distribution, telephony, and data transmission.

Nevertheless, this is not the only use to which fixed systems are put, and lower capacity systems are common for interconnecting commercial buildings that are separated by a few miles, or as a telemetry link relaying data back to a central computer location. Thus there exists a second layer of fixed links, both digital and analog. These do not form part of a national network and have a much lower capacity than the major system, however. Planning such links is carried out to the same standards as the high-capacity systems, though, and they are subject to the normal coordination procedures.

Thus these lower capacity systems appear to share many features with the modern mobile systems. They both operate in the microwave region of the radio spectrum, although not in the same band. They are usually fairly short and carry a mixture of telephony and data.

This being the case, it seems reasonable to expect the propagation-related aspects of fixed and mobile systems to have a high degree of commonality. As noted earlier in this introduction, this is not necessarily so, and we fully explore the underlying reasons for this in this chapter.

12.2 PATH LOSS

12.2.1 Fixed Links

Let us start by reexamining the target when planning a microwave LOS link and exploring the factors involved. We can define the target as "to attain free-space path loss between the system's antennas when no fading activity is present." To achieve this, we have to ensure that the radio path has 0.6 of the first Fresnel zone clear of any obstruction under conditions of severe subrefractivity, using suitable antenna heights at each end of the link. The degree of subrefractivity is defined as $k = 2/3$; that is, when calculating the height correction of obstacles to take account of both Earth and ray curvature, we use an effective Earth curvature of 2/3 of the true curvature in combination with a flat ray.

When we plan a link on this basis, the loss on the path between antennas is

$$L_{FS} = 32.5 + 20\log d + 20\log f \text{ dB} \tag{12.1}$$

where

f = the carrier frequency, in megahertz

d = the path length, in kilometers

Note that there is no diffraction loss involved.

12.2.2 Mobile Systems

If we first of all consider the VHF noncellular systems, the target is "to achieve the best possible coverage of the service area from a single base station." The obvious move puts the base station antenna as high as possible. This does not necessarily mean a high mast, as someone could use the high ground on which to place the support structure. Nevertheless, it is reasonable to expect that the higher the antenna relative to the terrain within the service area, the less likely the occurrence of black spots, because of the better look angle obtained.

Although the base station antenna would be placed as high as possible within the general restraints of overall system cost, the planner, unlike the fixed-link counterpart, has no control over the height of the antenna at the far end of the path (i.e., the mobile). If the terrain in which the mobile operates is hilly, then in certain locations the signal could only be received by a diffraction path with its associated path loss.

To give some feel for the severity of the diffraction loss, let us look at a rural location for which the distance of the mobile from the base station is 20 km, the effective height of the base station antenna is 150m, and the mobile antenna height is 1.5m. By using the curves in Figure 9.1, and applying a height-gain correction of –3 dB to convert the mobile antenna height from the 3m used within the curves to 1.5m, we reach a 50% of locations, 50% of the time, field strength 29 dB below the FS value.

The above example assumes that the service area closest to the one considered is sufficiently remote not to be considered from the point of view of cofrequency interference. If this were not the case, then the base station effective antenna height might have been restricted to avoid this problem. This would, of course, have resulted in an increase in diffraction loss.

So we learn that the path loss between a base station and a mobile can be very much greater than that between the antennas of a LOS link, due to diffraction losses. These losses arise from the fact that the mobile antenna height is very low so that the path-clearance criteria used for planning the point-to-point systems cannot be met for the majority of locations in a mobile situation. In the unusual situation whereby a mobile finds itself on high ground with a clear view of the base station antenna, then achieving FS path loss, but this is very much the exception rather than the rule.

12.3 MULTIPATH AND MEAN-DEPRESSION FADING

12.3.1 Fixed Links

For all but a small proportion of time, we can achieve the target of FS path loss between the antennas at the two ends of a microwave link. Under certain meteorological conditions, however, atmospheric layering occurs, resulting in abnormal rates of change in RRI over small height intervals. This can lead to transmitted energy possibly arriving at the receiver by more than one path because of reflection from or refraction within these layers. If the paths are of different physical length, then, because of the finite propagation velocity, the signals suffer different delays. Depending on the geometry of the situation at any one instant, the signals following the different paths may combine destructively to cause fading or, alternatively, in such a way as to cause signal enhancement. This ever-changing situation is often referred to as *multipath fading*. Examination of the fading patterns, however, reveals that in the vast majority of cases only two paths exist, so the term *multipath* is a misnomer.

Since the way in which signals combine is determined by their relative phase, it follows that for any given situation the phase difference between the signal components traveling over two different paths depends on the carrier frequency. At any one instant, then, some frequencies suffer a deep fade but others do not. Thus the alternative name for this fading—*frequency-selective fading*—is much more descriptive.

Examination of a record of a fading event (Fig. 2.1) reveals that besides the frequency-selective fading there is a second, slowly changing component of fading. This component is nonfrequency selective and is called the *flat-fading* or *mean-depression* term. It arises from such mechanisms as defocusing of the transmitted beam, sometimes called *beam-spreading,* or as a result of sub- or superrefractive layers causing the beam to reach off-axis angles and to suffer attenuation from the narrow beamwidth of the antennas used.

Layering in the atmosphere can cause two different types of fading on microwave LOS links. We now examine the types of fading encountered in the mobile environment.

12.3.2 Mobile Systems

Figure 8.3 represents the situation in which the mobile finds itself working. Assuming the mobile is at street level in an urban area, signal components arrive by many different routes. One thing that is certain is that there is very seldom a LOS path between the base station and the mobile. The signal arriving from the base station is a mixture of diffracted and reflected paths (or mixtures of both) all arriving with different transmission delays. The shortest route is the diffraction path over rooftops between base station and mobile, and the path loss involved changes in severity as the mobile moves along the street and the geometry slowly changes. We can liken this component to the mean-depression term of fixed-link fading.

What about the other signal components? They are obviously of different phase and amplitude having traveled different distances and suffered different mixes of reflection and diffraction. Thus we have a resultant that is the sum of a number of vectors having random phase and amplitude. This is the definition of *Rayleigh fading,* which is the term often mistakenly applied to the frequency-sensitive term of fading on LOS links. The fading on LOS links is in fact *Rician,* although the cumulative distributions of the two types of fading are very similar.

So we see that the fading on both services comprises a flat-fading term and a frequency-selective term, the main difference being that in the case of fixed links it results from atmospheric layering, a natural phenonemon and a time variant; whereas in the mobile situation, it is the man-made environment that is responsible and is position-dependent as the mobile moves along a street.

We must also consider an activity factor. Fixed links suffer deep fading for around 10^{-1}% of the month, and it is the depth of fading in the range 10^{-1}% to 10^{-4}% that is of interest to system planners and propagationists. In contrast, the fading in the mobile environment is always present, arising from the constantly changing relationship among the positions of the mobile, the diffracting, and the reflecting buildings and vehicles.

12.4 MULTIPATH DELAYS

12.4.1 Fixed Links

The multipath delays on microwave LOS links usually have an upper limit of 6 ns. This does not mean that longer delays do not exist but simply that any components having delays in more than this figure are small in amplitude and incapable of causing deep fading. This is because layering that causes delays greater than 6 ns must be considerably off the line of shoot of the antennas, and hence, any energy traversing such a path has significant offset launch and arrival angles with the associated main-lobe attenuation. Therefore, even if it arrives in antiphase to the direct signal, the degree of cancellation is only moderate. Thus it is the directivity of the narrow-beam antennas that restricts the effects of long-delay components.

12.4.2 Mobile Systems

In sharp contrast to the above, mobile systems have to contend with much longer multipath delays. GSM has an upper limit, for equalization, of 51 μs.

Again it is the antenna characteristics that play the major role—the type used in mobile systems having no significant directivity in the horizontal plane. Thus, delayed components reflected off distant objects suffer no attenuation introduced by the antennas, only path loss because of the distance covered.

12.5 DIVERSITY

12.5.1 Fixed Links

Multipath fading on fixed links is two-path in the majority of cases. Thus the vertical interference pattern shown at the receive end of a link, in Figure 3.1(a), is well defined and regular. It is not stationary since the atmospheric layering responsible for the multipath is neither static nor uniformly smooth. If the main and diversity antennas are separated by 150 to 200 wavelengths (center to center), then the fading patterns on the two antennas are virtually decorrelated. The signals from the two antennas can then be used in either the switched mode or the phase-aligned mode for effective diversity.

12.5.2 Mobile Systems

Fortunately, the 150- to 200-wavelength criterion does not apply to mobile systems. Because there are so many paths between base station and the mobile, with the relative phases changing rapidly as the mobile (and other traffic) moves, the distance required between antennas at the mobile to have a reasonable degree of decorrelation is less than one wavelength. Thus, at 950 MHz, a separation of 32 cm can give an effective diversity action. Such a distance obviously presents no problem on a vehicle but is not feasible on a handheld device.

In the case of the mobile-to-base station path, a horizontal separation of 11 to 13 wavelengths can give a correlation factor of 0.7, and the use of base station diversity is commonly.

12.6 FREQUENCY REUSE

12.6.1 Fixed Links

The high-capacity microwave links are in the 4-, 6-, 8-, 11-, and 19-GHz bands. Because the limited number of RF channels available, frequency reuse must be used. So, what precautions should you take to minimize the risk of cochannel interference?

First, the use of high-gain antennas to achieve the desired input level to the receiver without having to resort to high-power sources at the transmitter, automatically offers protection because of its inherent narrow beamwidth. A typical antenna beamwidth in such systems is 1° between the 3-dB points and, therefore, unless a cofrequency interfering signal is very closely aligned with the line of shoot of a link, antenna directivity results in a high-attenuation of the interfering signal. Thus, basic system characteristics together with good planning to avoid close angular alignment with existing links, offer effective protection against the problem.

However, there is always the possibility of a new link being planned that might cause a problem. Such a possibility is negated by the use of coordination procedures, both national and international, to ensure that not only is interference of this nature not possible under normal conditions, but nor is it a problem under ducting situations.

12.6.2 Mobile Systems

One important feature of cellular mobile systems is their high traffic-carrying capability caused by adoption of frequency reuse. There are several techniques to minimize the risk of interference from a cell into its nearest cofrequency neighbors. Among these are

- Siting the base station antennas as low as possible, while offering good radio coverage of its own service area;
- Using RF power control in the mobile and base station to reduce the transmit power to the minimum required to achieve the quality objective;
- Tilting the base station antennas downward to reduce spillover of transmitted power beyond the cell boundary.

When we consider the demands on the system in an urban situation, it is obvious that there are conflicting requirements. To handle the large number of users within a relatively small area, cell-splitting must be adopted, reducing the mean reuse distance, with its consequential threat of cochannel interference.

The use of sectorization, Figures 10.1 and 10.2, can reduce the threat by reducing the number of primary interferers capable of causing a problem. This happens by introducing some antenna directivity into the horizontal plane of the radiation pattern.

The introduction of digital systems has relieved the problem somewhat because they are more robust. Typically, the threshold carrier-to-interference ratio for GSM is 9.5 dB, compared with 17 dB for TACS equipment. Also, GSM uses discontinuous transmission and reception, under the control of a voice-activity detector, which reduces the cochannel interference by 3 dB, on average.

12.7 SUMMARY

We discovered that, when comparing propagation effects on fixed and mobile links, you may not only have to account for the particular propagation aspect but also for the system environment and, at times, the antenna characteristics.

Examining the topic of path loss, for instance, fixed links are planned in such a way that ground effects are avoided, even under difficult conditions. This is achieved by planning for very subrefractive situations ($k = 2/3$) and, by careful choice of antenna heights, by ensuring that the height of the radio path over obstacles allows for the correct Fresnel-zone clearance. This approach leads to FS path loss between the antennas, a situation that you cannot improve.

The mobile situation is very different, however, with the planner not being allowed the same freedom of choice in relation to antenna heights because the height of the mobile's antenna is, of course, very much restricted. Thus the path loss between base station and mobile must, in virtually all cases, include an allowance for diffraction loss.

Multipath fading and mean depression are in both the fixed-link and mobile environments. However, in the case of fixed links, excess path loss from these effects is caused by atmospheric layering, a natural phenomenon present for only a small percentage of time. Further, in multipath fading, the high directivity of the antennas used rules against a multiplicity of delayed components, two-path situations being the normal case.

For mobiles, however, multipath fading is caused by the interaction between a large number of delayed components, resulting from diffraction around edges and reflection from surfaces of buildings, vehicles, and so forth. Thus the manmade environment in which the mobile finds itself results in a very complex multipath situation that is ever present and changes rapidly as the mobile changes its position. The whole problem is enhanced by the fact that there is no antenna directivity effect to reduce the number of components.

The shadowing term (equivalent to the mean depression in fixed-link systems) is the diffraction loss experienced by the most direct signal component between base station and mobile as it crosses the rooftops. This loss varies as the mobile moves along the street, although the rate of change is much slower than that experienced by the multipath term as the relative phase of the many component signals alters.

Multipath delays are of particular interest in the specification of digital mobile equipment as they determine the range of equalization that must be provided. GSM caters for a maximum delay of 51μs, based on results from a number of measurement programs. This is a practical value, representing the sort of delay that can be experienced in reflections from nearby hills or from buildings in a large town. Fixed links, however, are unlikely to experience multipath components of large amplitude, having delays greater than 6 ns, due to the highly directional antennas rejecting off-axis signals, and the high back-to-front ratios of the antennas rejecting reflected signals coming from behind the receiving antenna. Thus we once again have a great difference between fixed links and mobile systems as a result of the differing characteristics of the antennas.

When considering the logistics of diversity, understand that with fixed systems, the vertical interference pattern at the receive end of a link suffering two-path fading is regular although not necessarily stationary. With antennas spaced by 150 to 200 wavelengths, the fading patterns on the two antennas are sensibly decorrelated, and good diversity action is possible. In the mobile environment, however, the large number of rapidly changing multipath components present creates a very complex interference pattern, and it has been shown that at the mobile an antenna separation of one wavelength can yield an effective diversity action. Nevertheless, such a spacing rules out diversity on handheld devices. At the base station, however, the antennas are in a relatively clutter-free environment, requiring a larger antenna separation.

Mobile systems owe their whole existence to the ability to reuse frequencies within a reasonably short distance. This points to the necessity to plan the system very carefully to avoid unnecessary spillover into cochannel cells. Countermeasures include, siting the antenna as low as possible, while taking care for good illumination of its own service area; tilting the base station antenna downward to reduce spillover at the cell boundary and employing sectorization. Countermeasures do not have to be restricted, however, to the antennas. The use of power control in the mobile and base station reduce the cochannel interference, as does the use of more robust equipment with a low-threshold carrier-to-interference ratio. GSM is 7.5 dB less sensitive to interference than AMPS.

Fixed links also reuse channel frequencies time and time again. The density of the network, however, is far less than that of the mobile systems. Also, the directivity of the antennas (1-degree beamwidth) offers high rejection of signals arriving off axis. The most important factors in the battle against cofrequency interference, however, are the coordination procedures, both nationally and internationally, to ensure at the planning stage that no proposed new links will pose a problem for the existing networks even under long-distance ducting conditions.

▼▼▼

About the Author

John Doble joined the Post Office Research Station (now BT Labs) straight from school. He studied part time for a degree in electrical engineering, graduating from London University in 1960.

He has been involved in many aspects of the development of high-capacity microwave radio systems, both analog and digital, and was responsible for the design of the automatic satellite tracking system on the first antenna at Goonhilly Earth Station.

In 1977, he was appointed Head of Terrestrial Propagation Studies Group and was heavily involved in the identification of propagation problems associated with the changeover of high-capacity routes from analog to digital working. As a result of his work in this area he became a member of the U.K. delegation to the CCIR over the period 1978–1992.

He retired from BT Labs in 1987 and took up a consultancy position with the Radiocommunications Agency (the radio-regulatory authority in the United Kingdom), specializing in mobile communications. In connection with this work, he was appointed chairman of the U.K. Mobile Radio Propagation Technical Working Party, became a member of the National Radio Propagation Committee, and was appointed U.K. Member to COST 231.

He retired from his consultancy position in 1992 but continues his work as a visiting lecturer at a number of universities, teaching line of sight and mobile propagation to MSC students. He also teaches the same subjects at short courses in the United Kingdom, Singapore, Indonesia, and Australia.

▼▼▼

INDEX

The Artech House Telecommunications Library

Vinton G. Cerf, Series Editor

Access Networks: Technology and V5 Interfacing, Alex Gillespie

Advanced High-Frequency Radio Communications, Eric E. Johnson, Robert I. Desourdis, Jr., et al.

Advanced Technology for Road Transport: IVHS and ATT, Ian Catling, editor

Advances in Computer Systems Security, Vol. 3, Rein Turn, editor

Advances in Telecommunications Networks, William S. Lee and Derrick C. Brown

Advances in Transport Network Technologies: Photonics Networks, ATM, and SDH, Ken-ichi Sato

An Introduction to International Telecommunications Law, Charles H. Kennedy and M. Veronica Pastor

Asynchronous Transfer Mode Networks: Performance Issues, Second Edition, Raif O. Onvural

ATM Switches, Edwin R. Coover

ATM Switching Systems, Thomas M. Chen and Stephen S. Liu

Broadband: Business Services, Technologies, and Strategic Impact, David Wright

Broadband Network Analysis and Design, Daniel Minoli

Broadband Telecommunications Technology, Byeong Lee, Minho Kang and Jonghee Lee

Cellular Mobile Systems Engineering, Saleh Faruque

Cellular Radio: Analog and Digital Systems, Asha Mehrotra

Cellular Radio: Performance Engineering, Asha Mehrotra

Cellular Radio Systems, D. M. Balston and R. C. V. Macario, editors

CDMA for Wireless Personal Communications, Ramjee Prasad

Client/Server Computing: Architecture, Applications, and Distributed Systems Management, Bruce Elbert and Bobby Martyna

Communication and Computing for Distributed Multimedia Systems, Guojun Lu

Community Networks: Lessons from Blacksburg, Virginia, Andrew Cohill and Andrea Kavanaugh, editors

Computer Networks: Architecture, Protocols, and Software, John Y. Hsu

Computer Mediated Communications: Multimedia Applications, Rob Walters

Computer Telephone Integration, Rob Walters

Convolutional Coding: Fundamentals and Applications, Charles Lee

Corporate Networks: The Strategic Use of Telecommunications, Thomas Valovic

Digital Beamforming in Wireless Communications, John Litva and Titus Kwok-Yeung Lo

Digital Cellular Radio, George Calhoun

Digital Hardware Testing: Transistor-Level Fault Modeling and Testing, Rochit Rajsuman, editor

Digital Switching Control Architectures, Giuseppe Fantauzzi

Digital Video Communications, Martyn J. Riley and Iain E. G. Richardson

Distributed Multimedia Through Broadband Communications Services, Daniel Minoli and Robert Keinath

Distance Learning Technology and Applications, Daniel Minoli

EDI Security, Control, and Audit, Albert J. Marcella and Sally Chen

Electronic Mail, Jacob Palme

Enterprise Networking: Fractional T1 to SONET, Frame Relay to BISDN, Daniel Minoli

Expert Systems Applications in Integrated Network Management, E. C. Ericson, L. T. Ericson and D. Minoli, editors

FAX: Digital Facsimile Technology and Applications, Second Edition, Dennis Bodson, Kenneth McConnell and Richard Schaphorst

FDDI and FDDI-II: Architecture, Protocols, and Performance, Bernhard Albert and Anura P. Jayasumana

Fiber Network Service Survivability, Tsong-Ho Wu

Future Codes: Essays in Advanced Computer Technology and the Law, Curtis E. A. Karnow

A Guide to the TCP/IP Protocol Suite, Floyd Wilder

Implementing EDI, Mike Hendry

Implementing X.400 and X.500: The PP and QUIPU Systems, Steve Kille

Inbound Call Centers: Design, Implementation, and Management, Robert A. Gable

Information Superhighways Revisited: The Economics of Multimedia, Bruce Egan

Integrated Broadband Networks, Amit Bhargava

International Telecommunications Management, Bruce R. Elbert

International Telecommunication Standards Organizations, Andrew Macpherson

Internetworking LANs: Operation, Design, and Management, Robert Davidson and Nathan Muller

Introduction to Document Image Processing Techniques, Ronald G. Matteson

Introduction to Error-Correcting Codes, Michael Purser

An Introduction to GSM, Siegmund Redl, Matthias K. Weber and Malcom W. Oliphant

Introduction to Radio Propagation for Fixed and Mobile Communications, John Doble

Introduction to Satellite Communication, Bruce R. Elbert

Introduction to T1/T3 Networking, Regis J. (Bud) Bates

Introduction to Telephones and Telephone Systems, Second Edition, A. Michael Noll

Introduction to X.400, Cemil Betanov

LAN, ATM, and LAN Emulation Technologies, Daniel Minoli and Anthony Alles

Land-Mobile Radio System Engineering, Garry C. Hess

LAN/WAN Optimization Techniques, Harrell Van Norman

LANs to WANs: Network Management in the 1990s, Nathan J. Muller and Robert P. Davidson

Minimum Risk Strategy for Acquiring Communications Equipment and Services, Nathan J. Muller

Mobile Antenna Systems Handbook, Kyohei Fujimoto and J.R. James, editors

Mobile Communications in the U.S. and Europe: Regulation, Technology, and Markets, Michael Paetsch

Mobile Data Communications Systems, Peter Wong and David Britland

Mobile Information Systems, John Walker

Networking Strategies for Information Technology, Bruce Elbert

Packet Switching Evolution from Narrowband to Broadband ISDN, M. Smouts

Packet Video: Modeling and Signal Processing, Naohisa Ohta

Personal Communication Networks: Practical Implementation, Alan Hadden

Personal Communication Systems and Technologies, John Gardiner and Barry West, editors

Practical Computer Network Security, Mike Hendry

Principles of Secure Communication Systems, Second Edition, Don J. Torrieri

Principles of Signaling for Cell Relay and Frame Relay, Daniel Minoli and George Dobrowski

Principles of Signals and Systems: Deterministic Signals, B. Picinbono

Private Telecommunication Networks, Bruce Elbert

Radio-Relay Systems, Anton A. Huurdeman

RF and Microwave Circuit Design for Wireless Communications, Lawrence E. Larson

The Satellite Communication Applications Handbook, Bruce R. Elbert

Secure Data Networking, Michael Purser

Service Management in Computing and Telecommunications, Richard Hallows

Smart Cards, José Manuel Otón and José Luis Zoreda

Smart Highways, Smart Cars, Richard Whelan

Super-High-Definition Images: Beyond HDTV, Naohisa Ohta, Sadayasu Ono and Tomonori Aoyama

Television Technology: Fundamentals and Future Prospects, A. Michael Noll

Telecommunications Technology Handbook, Daniel Minoli

Telecommuting, Osman Eldib and Daniel Minoli

Telemetry Systems Design, Frank Carden

Teletraffic Technologies in ATM Networks, Hiroshi Saito

Toll-Free Services: A Complete Guide to Design, Implementation, and Management, Robert A. Gable

Transmission Networking: SONET and the SDH, Mike Sexton and Andy Reid

Troposcatter Radio Links, G. Roda

Understanding Emerging Network Services, Pricing, and Regulation, Leo A. Wrobel and Eddie M. Pope

Understanding GPS: Principles and Applications, Elliot D. Kaplan, editor

Understanding Networking Technology: Concepts, Terms and Trends, Mark Norris

UNIX Internetworking, Second Edition, Uday O. Pabrai

Videoconferencing and Videotelephony: Technology and Standards, Richard Schaphorst

Wireless Access and the Local Telephone Network, George Calhoun

Wireless Communications in Developing Countries: Cellular and Satellite Systems, Rachael E. Schwartz

Wireless Communications for Intelligent Transportation Systems, Scott D. Elliot and Daniel J. Dailey

Wireless Data Networking, Nathan J. Muller

Wireless LAN Systems, A. Santamaría and F. J. López-Hernández

Wireless: The Revolution in Personal Telecommunications, Ira Brodsky

Writing Disaster Recovery Plans for Telecommunications Networks and LANs, Leo A. Wrobel

X Window System User's Guide, Uday O. Pabrai

For further information on these and other Artech House titles, contact:

Artech House
685 Canton Street
Norwood, MA 02062
617-769-9750
Fax: 617-769-6334
Telex: 951-659
email: artech@artech-house.com

Artech House
Portland House, Stag Place
London SW1E 5XA England
+44 (0) 171-973-8077
Fax: +44 (0) 171-630-0166
Telex: 951-659
email: artech-uk@artech-house.com

WWW: http://www.artech-house.com